Fullerenes

Synthesis, Properties, and Chemistry of Large Carbon Clusters

George S. Hammond, EDITOR
Bowling Green State University

Valerie J. Kuck, EDITOR
AT&T Bell Laboratories

Developed from a symposium sponsored
by the Divisions of Inorganic Chemistry; Organic Chemistry;
Petroleum Chemistry, Inc.; Polymer Chemistry, Inc.;
and Polymeric Materials: Science and Engineering
at the 201st National Meeting
of the American Chemical Society,
Atlanta, Georgia,
April 14–19, 1991

American Chemical Society, Washington, DC 1992

Library of Congress Cataloging-in-Publication Data

Fullerenes: synthesis, properties, and chemistry of large carbon clusters /
   George S. Hammond, editor, Valerie J. Kuck, editor

   p.   cm.—(ACS Symposium Series, ISSN 0097–6156; 481).

"Developed from a symposium sponsored by the Divisions of
Inorganic Chemistry, Organic Chemistry, Petroleum Chemistry, Inc.,
Polymer Chemistry, Inc., and Polymeric Materials: Science and
Engineering at the 201st National Meeting of the American Chemical
Society, Atlanta, Georgia, April 14–19, 1991."

   Includes bibliographical references and indexes.

   ISBN 0–8412–2182–0

   1. Fullerenes—Congresses. 2. Buckminsterfullerene—Congresses.

I. Hammond, George Sims, 1921–   . II. Kuck, Valerie J., 1939–   .
III. American Chemical Society. Division of Inorganic Chemistry.
IV. American Chemical Society. Meeting (201st: 1991: Atlanta, Ga.)
V. Series.

QD181.C1F85   1992
546'.681—dc20   91–40807
                                  CIP

The paper used in this publication meets the minimum requirements of American National
Standard for Information Sciences—Permanence of Paper for Printed Library Materials, ANSI
Z39.48–1984.   ∞

Copyright © 1992

American Chemical Society

ACS Symposium Series

M. Joan Comstock, *Series Editor*

1992 ACS Books Advisory Board

Foreword

THE ACS SYMPOSIUM SERIES was founded in 1974 to provide a medium for publishing symposia quickly in book form. The format of the Series parallels that of the continuing ADVANCES IN CHEMISTRY SERIES except that, in order to save time, the papers are not typeset, but are reproduced as they are submitted by the authors in camera-ready form. Papers are reviewed under the supervision of the editors with the assistance of the Advisory Board and are selected to maintain the integrity of the symposia. Both reviews and reports of research are acceptable, because symposia may embrace both types of presentation. However, verbatim reproductions of previously published papers are not accepted.

Contents

Preface vii

Overview ix
George S. Hammond

1. **Synthesis of C_{60} from Small Carbon Clusters: A Model Based on Experiment and Theory** 1
James R. Heath

2. **Crystalline Fullerenes: Round Pegs in Square Holes** 25
R. M. Fleming, B. Hessen, T. Siegrist, A. R. Kortan,
P. Marsh, R. Tycko, G. Dabbagh, and R. C. Haddon

3. **Low-Resolution Single-Crystal X-ray Structure of Solvated Fullerenes and Spectroscopy and Electronic Structure of Their Monoanions** 41
Sergiu M. Gorun, Mark A. Greaney, Victor W. Day,
Cynthia S. Day, Roger M. Upton, and Clive E. Briant

4. **Solid C_{60}: Structure, Bonding, Defects, and Intercalation** 55
John E. Fischer, Paul A. Heiney, David E. Luzzi,
and David E. Cox

5. **Conductivity and Superconductivity in Alkali Metal Doped C_{60}** 71
R. C. Haddon, A. F. Hebard, M. J. Rosseinsky, D. W. Murphy,
S. H. Glarum, T. T. M. Palstra, A. P. Ramirez, S. J. Duclos,
R. M. Fleming, T. Siegrist, and R. Tycko

6. **Crystal Structure of Osmylated C_{60}: Confirmation of the Soccer-Ball Framework** 91
Joel M. Hawkins, Axel Meyer, Timothy A. Lewis,
and Stefan Loren

7. **One- and Two-Dimensional NMR Studies: C_{60} and C_{70} in Solution and in the Solid State** 107
Robert D. Johnson, Costantino S. Yannoni, Jesse R. Salem,
Gerard Meijer, and Donald S. Bethune

8. **Mass Spectrometric, Thermal, and Separation Studies of Fullerenes** 117
    Donald M. Cox, Rexford D. Sherwood, Paul Tindall,
    Kathleen M. Creegan, William Anderson,
    and David J. Martella

9. **Production, Mass Spectrometry, and Thermal Properties of Fullerenes** 127
    Ripudaman Malhotra, Donald C. Lorents, Young K. Bae,
    Christopher H. Becker, Doris S. Tse, Leonard E. Jusinski,
    and Eric D. Wachsman

10. **Doping the Fullerenes** 141
    R. E. Smalley

11. **Survey of Chemical Reactivity of C_{60}, Electrophile and Dieno–polarophile Par Excellence** 161
    F. Wudl, A. Hirsch, K. C. Khemani, T. Suzuki, P.-M. Allemand,
    A. Koch, H. Eckert, G. Srdanov, and H. M. Webb

12. **The Chemical Nature of C_{60} as Revealed by the Synthesis of Metal Complexes** 177
    Paul J. Fagan, Joseph C. Calabrese, and Brian Malone

Author Index 187

Affiliation Index 187

Subject Index 188

Preface

In 1990, WHEN A COUPLE OF ASTROPHYSICISTS, W. Krätschmer and D. R. Huffman, and their co-workers reported the isolation from graphite soot of stable molecules having the compositions C_{60} and C_{70}, an avalanche of chemical research on these and related compounds was launched. The result was not a total surprise, because chemists (R. E. Smalley, H. W. Kroto, and co-workers) had already reported the extraordinary production and stability of these compounds in the vapor phase and proposed what are now their accepted structures. However, the work of Krätschmer and Huffman made the compounds available in macroscopic amounts and thereby made possible of the study of their physical and chemical properties by the myriad means available to chemical scientists.

This volume is the output from a special Fast Breaking Events Symposium held in April 1991 at the American Chemical Society National Meeting in Atlanta, Georgia. The first such symposium was presented at the ACS 1987 Spring Meeting in Denver, Colorado; it dealt with the then-new topic of high-temperature superconductivity. A year later in Dallas, a second special symposium was devoted to cold fusion.

The fact that the scientific follow-on from the first two special events symposia have been of rather different character is entirely consistent with ACS philosophy of providing special fora for these presentations. It is not expected that the symposia will be in any way final or definitive. They attempt to lay before the members of the society the results and ideas of researchers who are currently doing work that, because of its novelty and possible import, is of great interest to chemical scientists.

By the time we decided to organize the special symposium in December 1990, dozens of reports on the subject had appeared in the literature or were known in preprinted form. By the time of the symposium in April, the number had multiplied several fold. Now the number is in the hundreds. Development of the chemistry of the fullerenes certainly qualifies as a "fast breaking event"!

Acknowledgments

Unlike the first two special events symposia, this symposium was not a Presidential Event. We are grateful to the ACS Committee on Science and its chairman, Ivan Legg, for providing the financial support for this symposium. We thank the reviewers, who were most cooperative and reviewed each chapter thoroughly despite the short deadline.

We thank the divisions who cosponsored this symposium. In particular, we are pleased to note that the Organic Division, the Petroleum Division, the Polymer Division, and the Division of Polymeric Materials: Science and Engineering have donated their portions of the royalties from this book to Project Seed, a career development activity administered by the ACS. Project Seed funds economically disadvantaged high school students for 8–10 weeks to do research in an industrial or academic laboratory.

Finally, we thank the ACS Books Department staff, especially A. Maureen Rouhi, for their assistance in preparing this book and publishing it in a timely manner.

GEORGE S. HAMMOND
P.O. Box 207
Painted Post, NY 14870

VALERIE J. KUCK
AT&T Bell Laboratories
Murray Hill, NJ 07974

September 1991

The Fullerenes: Overview 1991

George S. Hammond

THE APPEARANCE OF NEW ALLOTROPIC FORMS of elemental carbon has evoked surprise and excitement in chemists (and physicists). People have for many years speculated about the possible existence of allotropes other than diamond and graphite, but attention has been focused on other conceivable extended structures. To our knowledge, little thought was given to stable molecular structures before the detection by mass spectrometry of very large amounts of large molecules, especially C_{60} and C_{70}, containing even numbers of carbon atoms in certain carbon vapors[1]. These reports generated considerable interest and led to the hypothesis that the abundant mass peaks were due to molecules having relatively great stability. However, few were inclined to "count" the species as carbon allotropes. The new species, like small molecules such as C_2 and C_3, were thought of as transients in dilute vapor phases, hardly what one classifies as "materials" based on elemental carbon.

All of this changed dramatically when Krätschmer, Fostiropoulos, and Huffman reported[2] that macroscopic amounts of the compound C_{60}, along with mixtures of other C_{2n} species, could be isolated from soots produced by vaporization of graphite. The "reality" of the new series of compounds was clearly demonstrated. That reality was relatively easily accepted, partly because Kroto, Smalley, and their co-workers had already developed a theory based upon topological considerations to account for the special stability of C_{60} and C_{70}. Chemists have an instinctive delight in symmetry, and the proposed structures, which, in outline form, do look like soccer balls, not only provide tentative rationalization of the stability of the molecules but also have great aesthetic appeal. Half of the outlined stable structure is also reminiscent of geodesic domes, architectural structures of great stability and, to some, of great beauty developed by the late Buckminster Fuller. That relationship generated what has become the generic name, fullerenes, of the series of stable compounds. The same line of thought has led to playful designation of the molecules as "bucky balls."

The availability of relatively easily prepared and purified samples of C_{60} has started an avalanche of work around the world designed to elucidate the structure of the compound and its analogues. Essentially the entire arsenal of available tools for structural studies has been brought to bear on the problem. As well as providing conclusive information about the molecular structures, the work has illustrated the enormous power of the methods.

The ^{13}C NMR spectrum of C_{60} in solution shows a single, sharp resonance, indicating the equivalence of all carbon atoms in the molecules. The result is in perfect agreement with the proposal that C_{60} is a hollow molecule with the shape of a truncated icosahedron; in fact, I cannot conceive of any other way of placing 60 equivalent points in a three-dimensional array. The chemical shift of the resonance is similar to that of ^{13}C in torsionally strained aromatic hydrocarbons in good agreement with evolving theory of the electronic structures of fullerenes. The NMR spectra of C_{70} and other less symmetrical members of the group of compounds provide ancillary structural information that complements, and in no way contradicts, the conclusions concerning C_{60}.

X-ray diffraction from solid C_{60} provides stunning confirmation of the high symmetry of the molecules, but, at the same time, the molecular symmetry contrives to deny crystallographers immediate access to structural details at the atomic level of resolution. At ambient laboratory temperatures diffraction data cannot be refined to reflect the anticipated icosahedral symmetry. This condition implies that the molecules "hop" rapidly between symmetry positions, and this action makes them appear essentially as spheres to X-rays. The result is agreeably compatible with NMR data, which give single resonances, just as in solution, indicating that the crystal structure imposes no inequivalence of carbon atoms on the NMR time scale. Thermal data show that a phase transition (from face-centered cubic to primitive cubic) occurs at 249 K. Both magnetic resonance and X-ray data indicate that at low temperatures the molecules become symmetry-inequivalent. The details, which have not yet been entirely resolved, are discussed with great clarity in some of the chapters in this volume (e.g., Chapters 2–5).

X-ray diffraction by crystalline derivatives of C_{60} gives significantly enhanced structural information. Both heavy metal derivatives and alkali metal intercalates have been studied. Among the latter the potassium derivatives have been accorded the most attention. The superconducting K_3C_{60} exists as disordered crystals apparently derived directly from the parent body-centered cubic (bcc) structure. However, crystals that have been intercalated to saturation with alkali metals, reaching the composition M_6C_{60}, have a modestly modified crystal structure (bcc) and finally show the carbon cages "pinned" in a structure that reveals the clusters as having less than spherical symmetry. All C_{60} anions are completely oriented with respect to the crystal axes. It is easy to attribute the rigid orientation to electrostatic interaction between the C_{60}^{6-} units and the M^+ cations, although other explanations can be entertained. At any rate, if all C–C distances are constrained to be equal, the data indicate that distance to be 1.44 Å , slightly greater than the C–C bond lengths adduced from NMR data.

Addition of heavy metal groups to the fullerenes generates compounds in which crystal packing is partially controlled by adducted groups.

Hawkins and co-workers (Chapter 6) at Berkeley have prepared an osmium derivative of C_{60}, $[C_{60}(Os_6O_4)(4\text{-}t\text{-butylpyridine})_2]$ and have subjected it to single-crystal X-ray diffraction analysis. A Du Pont group (Chapter 12) has done the same with a platinum derivative $[(Ph_3P)_2Pt(N_2C_{60})$ Ď $(C_4H_8O)]$. In both cases the diffraction patterns can be unequivocally interpreted and directly reveal the icosahedral structure of the parent molecule with distortion, as would be expected, by the substituent metal atoms. The C–C bond lengths fall into two groups, as expected. The bonds for fusions of two six-membered rings are close to 1.39 Å , and those for fusions of a six- and a five-membered ring are near 1.43 Å . (The presence of the substituents introduces inequivalence in bond lengths, so the cited values are averages of *similarly* sited bonds.)

Optical spectra of C_{60} and C_{70} are informative and consistent with structural assignments. They show a series of bands in the visible and ultraviolet spectral regions. Although the spectra do not reveal the molecular structures in any simple way, they show that the molecules contain well-defined HOMOs and LUMOs, a feature that demands that the molecules have definitely fixed structures. The electronic spectra are, of course, valuable in development of theory of the electronic structures of the molecules.

The vibrational spectra of C_{60} are highly revealing. The infrared shows only four bands, and there are eight clearly developed Raman bands. There is no overlap between the bands in the two spectra, as is expected from a molecule of very high symmetry. Theory predicts that there should be four IR and 10 Raman bands. With a little imagination two additional Raman lines may be seen, so theory is stunningly successful!

The durable integrity of the fullerene structures suggests that they should undergo extensive chemical modification without disruption of their basic structures; and such is certainly proving to be the case. C_{60} has been most extensively studied to date. The compound has relatively favorable electron affinity and undergoes electrolytic one- and two-electron reversible reduction in solution to form fulleride (buckide) anions. Birch reduction by alkali metals in protic solvents yields highly hydrided species, for example, $C_{60}H_{36}$.

As has been mentioned earlier, solid C_{60} can be intercalated with vapors of alkali metals to produce materials having compositions M_xC_{60} where $x = 1\text{–}6$. The first metallic derivative of fullerene, LaC_{60}, was actually produced and detected by mass spectrometry in the very early days of study of the gaseous products from vaporization of graphite. That work and later study by the Rice group to establish the structure of the compound and to produce ponderable amounts of it are described by Smalley in his chapter in this volume. That the structure of the lanthanum derivative is very different from those of the metal intercalates is clearly demonstrated by its stability as a monomeric gaseous species. That it is uniquely produced by volatilization of graphite impregnated with $LaCl_3$ or La_2O_3 is consistent with the view that the metal atom is incorporated *inside* the cage. It has not been possible to produce the material by penetration into the preformed fullerene, so it appears that the metal atom must be bound to a partially formed carbon aggregate and that the rest of the cage is assembled around it.

Unlike metal derivatives in which metal atoms are held on the outside of the metal cages, the lanthanum compound cannot be easily dissociated either thermally or photochemically. On pyrolysis it loses C_2 units successively until it reaches the size $C_{42}La$, when it "blows"; the minimum critical size for metal-free fullerenes is C_{32}. Smalley has suggested a systematic formulation to distinguish between "inside" and "outside" fullerides; thus the formula $(La@C_{60})$ represents a lanthanum fulleride with the metal atom inside the cage. An outside compound, such as tripotassium fulleride, is represented by an ordinary linear formula, K_3C_{60} (or $C_{60}K_3$).

It is logical to ask, "What will come of it all?" and even, "What will fullerenes be good for?" The answers to such questions must be speculative, but current interest in fullerenes is so great that futuristic speculation seems warranted.

The existence and stability of the fullerenes, and their first noted properties, have already stimulated significant extension of electronic theory of molecular structure. That theory and extensions thereof will probably predict properties of related substances and provide guidance to synthetic work with fullerene derivatives.

It is impossible to divine which unexpected and valuable properties may be encountered as the number of fullerenes increases and sufficient amounts of material become available for determination of their macroscopic properties. For example, it does not a priori seem likely that fullerenes will have notable pharmacological properties. However, it is inevitable that sooner or later water-soluble derivatives will be subjected to biological screening; after all, few people predicted that aminoadamantane would be a potent antiviral agent. The speculation has appeared in the public press and elsewhere that fullerenes may be effective solid lubricants. The idea has some merit. Because of the absence of strong intermolecular interactions, the crystals are soft and will surely undergo easy deformation under shearing stress in a manner somewhat analogous to the way in which graphite functions as a lubricant because of easy slippage of the planar sheets in the crystals with respect to one another.

Although lubrication is not a glamorous aspect of materials science, it is representative of the fact that when new substances appear, engineers almost automatically wonder how they would behave in structural materials. Fullerenes will be accorded that attention even though my intuition does not predict startling results.

The property of fullerene derivatives that has attracted the most attention to date is the electrical conductivity of certain doped fullerides. Solid C_{60} adds potassium and rubidium to form the series $C_{60}M_x$ where $x = 1-6$. The adducts with $x < 6$ are conductors; in potassium fullerides conductivity is maximizes with $C_{60}K_3$. Films of the material are not only conductors but become superconducting at low temperatures. Zero resistivity of films has been observed at 5 K, and study of the Meissner effect by a.c. magnetization of powders indicates that $T_c = 19.3$ K. The bulk material doped with rubidium to the nominal composition $C_{60}Rb_3$ shows small, variable Meissner fractions by microwave loss and flux exclusion with T_c of 25–30 K. At the time of writing new reports of superconductivity are filtering in from around the world. Dopants now include electron acceptors (iodine) as well as donors, and unconfirmed reports of T_c values as high as 50 K have been heard. Probably the surface has barely been scratched in this fascinating area.

Basic structural theory to describe the electrical properties of the fullerides seems straightforward. The molecular orbital description of C_{60} indicates the availability of three LUMOs that would be half filled in the $C_{60}M_3$ derivatives. The face-centered cubic crystal structure for solid C_{60} provides three interstitial sites per C_{60} unit, and these sites can accommodate the positive counterions in $C_{60}M_3$ materials. The principal challenge to theorists, in my opinion, is to effect a smooth transition from the description of molecular energy levels to formulation of the Fermi levels and conduction bonds in the extended structures. This effort seems to be progressing well and will surely be extended to description of the special mechanism of charge flow in superconducting states. The fact that the fullerides are the first recognized *isotropic molecular conductors* is of inherent technological significance and may also facilitate significant advance in the theory of extended charge flow in solids built from discrete molecular units.

The torrent of research publications related to the fullerenes and their derivatives assures us that any snap-shot picture of the field will be an anachronism before it can appear in print. Here I have attempted not a review, but an overview from the perspective of an *observer* in the field. I only hope that I have not been guilty of grave distortion of the work and views of *participants* in this evanescent science.

[1]Kroto, H. W.; Heath, J. R.; O'Brien, S. C.; Curl, R. F.; Smalley, R. E. *Nature (London)* **1985**, *318*, 162–163.

[2]Krätschmer, W.; Lamb, L. D.; Fostiropoulos, K.; Huffman, D. R. *Nature (London)* **1990**, *347*, 354.

Chapter 1

Synthesis of C_{60} from Small Carbon Clusters

A Model Based on Experiment and Theory

James R. Heath[1]

Department of Chemistry, University of California, Berkeley, CA 94720

A model based on experiment and ab initio theory for the high-yield carbon-arc synthesis of C_{60} and other fullerenes is presented. Evidence that is given indicates that the synthesis must start with the smallest units of carbon (atoms, dimers, etc.). The model is then broken into four steps: (1) the growth of carbon chains up to length C_{10} from initial reactants present in the carbon vapor, (2) growth from chains into monocyclic rings (C_{10}–C_{20}), (3) production and growth of three-dimensional reactive carbon networks (C_{21}–C_x, x = 30–40), and (4) growth of small fullerene cages via a closed-shell mechanism that exclusively produces C_{60}, C_{70}, and the higher fullerenes as the stable products.

The carbon-arc synthesis of C_{60} from solid graphite surely represents one of the most phenomenal phase transitions ever discovered. This synthetic technique, developed by Krätschmer, Fostiropoulos, and Huffman (*1*), has made bulk quantities of C_{60} and other fullerenes available to the scientific community. This availability has stimulated a tremendous effort directed at characterizing this novel and exciting new class of molecules. One of the most fascinating aspects of the fullerenes, however, is the chemistry of their formation within the carbon arc. The synthesis is amazingly simple, and yet it produces results that are so fantastic and unexpected! The best experimental evidence (more on this later) indicates that the graphite reactant is vaporized in the carbon arc into the smallest units of carbon—atoms and possibly dimers—which then, through a concerted series of reactions, and in a limited range of pressures and temperatures, recombine to produce the spheroidal shells of carbon known as the fullerenes (*2*).

How can this reaction scheme be understood? What are the individual chemical mechanisms that lead to the formation of the fullerenes? What reactive intermediates are produced in the carbon arc? How does the high-temperature chemical environment of the carbon arc produce the low-entropy

---

[1]Current address: IBM T. J. Watson Research Laboratories, P.O. Box 218, Yorktown Heights, NY 10598

0097–6156/92/0481–0001$06.00/0

C_{60} molecule in such high yield? Complete answers to this complex set of questions will undoubtedly take many years. It is possible, however, to begin to put together a crude model that draws from modern ab initio quantum techniques, new experiments on C_{60}, and recent experimental advances in the study of smaller carbon clusters. This chapter will attempt to present such a model.

The Carbon-Arc Synthesis: Determination of Initial Reactants

With a problem this complex, perhaps the best technique for constructing a working model is to initially back away from the problem and attempt to address only its most basic aspects. One such aspect involves the nature of the carbon species initially ejected from the graphite rod by the carbon arc. Is the carbon ejected as large pieces that somehow manage to fold up, expelling and/or adding smaller molecules of carbon, until a fullerene structure is created? Or, conversely, is the carbon ejected as atoms and/or very small molecules, which then build up to form the framework of a fullerene? These questions may be addressed through isotopic scrambling experiments. In these experiments, the starting material consists of regions of pure ^{13}C and regions of natural-abundance ^{13}C. If the isotopic distribution of a fullerene produced in this arc is statistical, and the ^{13}C atoms are randomly distributed throughout the C_{60} framework, then most likely a small-molecule synthesis is occurring. If, however, the dispersion of ^{13}C is nonstatistical, then the experiment is inconclusive. A nonstatistical result is interpretable as either a large-molecule formation mechanism or as an indicator that the gas-phase mixing of the ^{13}C and ^{12}C regions is too slow to compete with the kinetics of carbon condensation.

Isotopic scrambling experiments have been done by Meijer and Bethune (3) and by the C_{60} team at Berkeley. In the Berkeley experiment, a 0.25-inch-diameter graphite rod is modified by drilling three 1.25-inch-deep, 0.06-inch-diameter holes down the bore of the rod. These holes are then packed with 98% ^{13}C-enriched amorphous carbon (Isotec, Inc.). In this way, a 1.25-inch segment of a graphite rod is enriched with 10–15% by weight ^{13}C. This rod is then mounted into the Berkeley C_{60} synthesis apparatus, which is roughly modeled after the design of Smalley and co-workers (4). The terminal 1 inch of the enriched portion of the rod is vaporized in a 100-torr (13.3 × 10^3-Pa) atmosphere of argon in about 20 s, and the carbon soot material is collected. This material is thoroughly washed for 1 h with ether in a reflux column and subsequently extracted with benzene, using the same column, for 3–5 h, or until the benzene wash is clear. The benzene is evaporated, and the fullerene extract is collected and stored under vacuum or in an inert (argon) gas environment while it awaits further characterization. This technique produces approximately 60–100 mg of ^{13}C-enriched fullerene material, corresponding to roughly a 10% yield based on the graphite and amorphous ^{13}C starting materials.

Isotopically resolved mass spectra of electron-impact ionized ^{13}C-enriched C_{60} are shown in Figure 1 (top). If the ^{13}C has been randomly introduced into the fullerenes, then the dispersion of isotopomers should fit the pattern of a binomial distribution (5):

$$F(n,x) = \frac{p^x q^{n-x} n!}{x!(n-x)!} \tag{1}$$

for a $^{13}C_x^{12}C_{n-x}$ molecule. Here, $n = 60$, x is the number of ^{13}C atoms substituted into the molecular framework, p is the probability a given atom is ^{13}C, and q ($= 1 - p$) is the probability that the atom is ^{12}C. Initial analysis of the mass spectra indicate that the isotopic distribution is quite nonstatistical. Indeed, it is impossible to simulate it with a single binomial distribution. However, the spectra may be modeled reasonably well as a bimodal binomial distribution, and such a model is shown in the bottom half of Figure 1. This bimodal distribution indicates that most (two-thirds) of the $^{12}C-^{13}C$ vapor completely scrambles to produce a 7% ^{13}C enrichment, while one-third of the vapor is barely mixed at all, and, indeed, is very close (1.7%) to the 1% ^{13}C natural abundance.

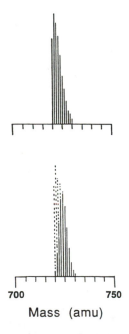

Mass (amu)

Figure 1. Top: Isotopic distribution of C$_{60}$ generated from a carbon rod that has been hollowed and filled with ^{13}C powder to give a 10% by weight ^{13}C rod. Bottom: Simulation of the experimental data using two separate binomial distributions. In this simulation, one-third of the C$_{60}$ is assumed to be 1.7% ^{13}C (solid line), and two-thirds of the carbon is assumed to be 7% ^{13}C (dashed line).

The bimodal distribution does have a physical basis if one considers the experimental technique for preparing this material. The three bore holes filled with ^{13}C are designed to "premix" the material within the rod prior to vaporization. However, the premixing is incomplete, and, indeed, much of the graphitic carbon is 1–2 mm away from the ^{13}C-filled holes. Thus, if fullerene synthesis occurs on a time scale shorter than the gas-phase mixing of the vaporized carbon, then only a portion of the gas-phase synthesis will be accomplished in a ^{13}C-enriched volume. This model may be tested by vaporizing a rod containing a similar weight-percent of ^{13}C, but now with only a single ^{13}C-filled hole, so that premixing within the rod is less complete. Mass analysis of C_{60} prepared in this way also indicates a roughly bimodal distribution of isotopomers, although now the amount of low-enrichment material accounts for more than 50% of the whole.

The isotopic scrambling experiments just discussed do provide a compelling, although not definitive, argument for a small-molecule synthesis of C_{60}. Complimentary evidence comes from the NMR experiments of Johnson et al. (6) performed on isotopically enriched C_{70}, prepared in a similar manner. These experiments indicate that the coupling between ^{13}C nuclei is no stronger than would be expected from a statistical dispersion of ^{13}C atoms throughout the C_{70} framework. This finding is expected only if the starting materials for synthesis of C_{70} are atoms or if the molecular framework of a very hot C_{70} molecule is actually fluxional, such that the carbon atoms can migrate from one position to another.

The mass spectrometric experiments, taken together with the NMR experiments, provide quite strong evidence that carbon-arc fullerene synthesis originates from atoms or, at most, from very small carbon clusters (dimers and trimers), but not from large graphitic sheets ejected from the carbon electrodes. With this fact established, it is now appropriate to view the synthesis of fullerenes a bit more closely. What is the chemistry of carbon vapor? What chemical intermediates are formed on the way to the fullerene synthesis? The rest of this chapter will be devoted to addressing these questions.

The Chemical Model

The initial reactant for production of the fullerenes is, apparently, the carbon atom (and possibly C_2), and the smallest stable fullerenes that, at this writing (September 1991), have been isolated are C_{60} and C_{70} (7). Thus, the number of unstable chemical intermediates potentially involved in carbon-arc fullerene synthesis is huge, and it includes all the structural isomers of the bare carbon clusters C_2 through C_{69}, excluding C_{60}. The structures of several of these clusters are either known from high-resolution spectroscopy, or they have been calculated by using sophisticated ab initio quantum chemistry techniques (8). A generalization of those structures is as follows: The smallest clusters (C_2 through C_9) are either known or predicted to have low-lying linear structures. Clusters in the range C_{10}–C_{20} are predicted (with some experimental support) to exist as a series of monocyclic rings. The next larger group, C_{21}–C_{31}, is difficult to characterize, and information concerning these clusters is scarce. They appear to consist of an unstable transition region between the monocyclic rings and the fullerene-type intermediates, which dominate the C_{32}–C_{69} size range. The odd-numbered clusters in the C_{32}–C_{69} size range are unstable toward loss of an atom (*9*), and are probably quite close to fullerene structures. Unlike even-numbered C_n, no closed solution contains only pentagons and hexagons for the odd-numbered clusters, however.

With these structures in hand, it is possible to present a crude picture of the various condensation steps of the carbon vapor, and such a picture is shown in Figure 2. Let's look a bit more closely at these various steps.

Step 1. Condensation of Very Small Carbon Species: Cluster Growth in One Dimension

Step 1 is essentially a "kickstart" of the carbon-condensation process. There are some chemical requirements for this step. First, it must favor the formation of pure carbon molecules and be biased against the addition of H, O, or N atoms. This requirement comes from experiments that show that bare carbon clusters are dominant combustion products in low-pressure acetylene–O_2 and benzene–O_2 flames (*10*). Carbon clusters are also formed in the laser ablation of polyimide (*11*).

Second, Step 1 must be extremely facile. A number of experiments, including the isotopic scrambling experiment discussed, indicate that condensation kinetics for carbon are remarkably fast—much faster, in fact, than for any other element. In general, when a vapor of atoms and dimers condenses to form larger clusters, the rate-limiting steps in cluster growth are always the initial ones. This condition prevails because atom–atom, dimer–atom, and dimer–dimer collisions require a third body to stabilize the collision complex, and three-body kinetic processes are notoriously slow. Carbon may provide the lone exception to this rule.

To understand this general rule, consider the physics of a collision complex. RRKM transition-state theory dictates that the lifetime of such a complex is purely statistical (*12*). The kinetic model assumed here is

$$C_2 + C \xrightarrow{k_b} C_3{}^* \xrightarrow{k_a{}^*} C_3{}^\ddagger \longrightarrow C_2 + C \qquad (2)$$

where $C_3{}^*$ is the energized molecule, $C_3{}^\ddagger$ is the activated complex, k_b is the collision rate, and $k_a{}^*$ is the rate of formation of the activated complex from the energized molecule. The internal energy of the energized molecule (E_v) contains several contributions:

$$E_v = D_e + E_c (= C_e + \mathrm{IE_r} - T_e) \qquad (3)$$

where D_e is the dissociation energy of the bound molecule, E_c is the energy available to the activated complex, C_e is the collision energy of the reactants,

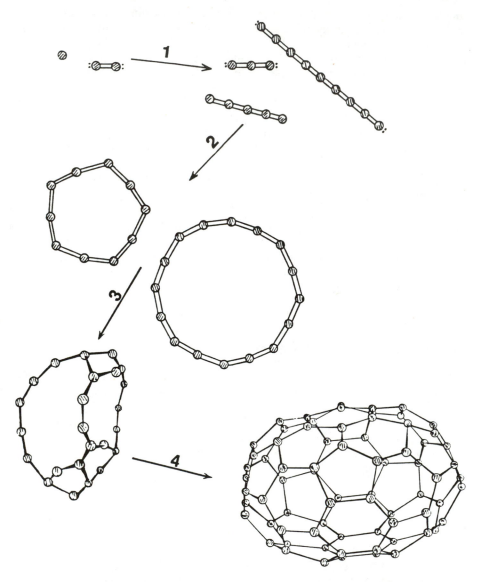

Figure 2. A chemical model for the generation of fullerenes from atomic and dimeric carbon vapor. Step 1 constitutes the initial condensation process—the formation of C_n chains (n < 10). Step 2 leads to the formation of monocyclic rings for C_n (9 < n < 21). Step 3 is the formation of unstable C_n (20 < n < 32), which quickly proceed through Step 4 to produce the fullerenes. Experimental and theoretical evidence for this condensation model is discussed in the text.

IE$_r$ is the contribution from the internal energy of the reactants, and T_e is the translation energy of the collision complex. E_c is partitioned into the rovibrational levels of the complex. Some of these levels will correspond to motion along the reaction coordinate, and they provide pathways for the activated complex to revert back to reactants. In a diatomic molecule, all available energy levels correspond to such motion, and thus a third body is always necessary to stabilize the complex, including the case of C + C. For a polyatomic molecule, the vibrational manifold is more complex, and a polyatomic molecule, with E_v above its dissociation threshold, has a finite, calculable lifetime. A calculation of the lifetime of C$_3$* gives great insight into why the condensation of carbon vapor is so facile.

Such a calculation requires knowledge of vibrational frequencies, and a list of measurements for C$_3$ and several other of the linear carbon clusters is presented in Table I. Ab initio predictions of all the harmonic vibrational frequencies for linear C$_n$ are listed in Table II. All of these clusters possess extremely low-frequency degenerate bending (π_g and π_u) modes. This is a key point, and will be important throughout this chapter. The ν_1, ν_2, and ν_3 vibrational frequencies of C$_3$ have been measured (13–16), and these frequencies provide input for a calculation of the lifetime of the [C$_2$ + C] collision complex from RRKM theory (17). A semiclassical estimation of the vibrational state density (N) of the energized molecule at E_v is (18)

$$N(E_v) = \frac{(E_v + E_z)^{s-1}}{(s-1)! \prod_{L=1}^{s} h\nu_i} \tag{4}$$

Here, E_z is the zero-point energy of the bound molecule ($\frac{1}{2}\Sigma h\nu_i$), $h\nu_i$ are the vibrational frequencies, and s is the number of normal modes. The number of vibrational levels, $P(E^*)$, available to the transition-state complex is

$$P(E^*) = \frac{(E_c + E_z)^s}{s! \sum_{i=1}^{s-1} h\nu_i} \tag{5}$$

where the summation is taken over all vibrational modes except the reaction coordinate (taken here to be ν_3). The lifetime of a [C$_2$ + C] collision may then be estimated from

$$\frac{1}{k(E^*)} = \left(\frac{Q}{Q^+}\right)^{1/2} \left[\frac{hN(E_v)}{P(E^*)}\right] \tag{6}$$

Here the $\frac{1}{2}$ accounts for the fact that dissociation of C$_3$ may occur at either end of the molecule. The factor (Q/Q^+) is a ratio of rotational partition functions for C$_3$* and C$_3^\ddagger$, and here it is set to 1 (adiabatic rotation approximation).

Table I. Molecular Constants for Linear C_n Obtained from Fluorescence, IR, and FIR Spectroscopy

C_n State	ν_o (cm^{-1})	B (cm^{-1})	D ($\times 10^5$ cm^{-1})	H ($\times 10^7$ cm^{-1})	Ref.
C_3 ($^1\Sigma_g^+$)		0.4305723 (56)	0.1472 (13)	0.1333 (59)	
$\nu_1 = 1$	1223 (3)	0.41985 (89)			15
$\nu_2 = 1^1$	63.416529 (40)	0.4424068 (52)[a]	0.2361 (16)[b]	0.267 (12)[c]	13
$\nu_3 = 1$	2040.0192 (6)	0.435704 (19)	0.4238 (31)	0.994 (23)	14
C_4 ($^3\Sigma_g^-$)		0.16452 (5)	N.D.[d]	0.0[e]	
$\nu_3 = 1$	1548.9368 (21)	0.164867 (7)	0.87 (19)	0.0	26
C_5 ($^1\Sigma_g^+$)		0.0853133 (29)	0.00053 (4)	0.0	
$\nu_3 = 1$	2169.4410 (2)	0.0848933 (29)	0.00053 (4)	0.0	29
$\nu_7 = 1^1$	118 (3)	0.0856235 (24)	0.00042 (4)	0.0	29
$\nu_7 = 2^0$	N.D.	0.085911 (7)	0.00093 (16)	0.0	29
$\nu_5 = 1^1$	218 (13)	0.085654 (4)	0.00082 (8)	0.0	29
C_7 ($^1\Sigma_g^+$)		0.030613 (14)	−0.00233 (85)	−0.0054 (15)	
$\nu_4 = 1$	2138.3152 (5)	0.030496 (14)	−0.00251 (91)	−0.0059 (17)	31
$\nu_5 = 1$	1898.3758 (8)	0.030556 (15)	−0.0016 (11)	−0.0044 (31)	32
$\nu_{11} = 1^1$	N.D.	0.03358 (40)	0.0	0.0	31
$\nu_{11} = 2^0$	N.D.	0.0374 (30)	0.0	0.0	31
C_9 ($^1\Sigma_g^+$)		0.014319 (16)	0.0014 (6)	N.D.	
$\nu_6 = 1$	2014.3383	0.014286 (15)	0.0011 (5)	−0.00037 (9)	33

NOTE: The numbers in parentheses are statistical uncertainties.

[a] $q_l = 0.0056939$ (21).

[b] $q_D = -0.0869$ (27).

[c] $q_H = 0.027$ (23).

[d] N.D. indicates not determined.

[e] A value of 0.0 indicates that the constant was set to 0.0 for the fit.

The lifetimes of the collision complexes for the reactions [C_2 + C], [C_2 + O], and [C_2 + H], as a function of E_c, are shown in Figure 3. These calculations employed vibrational basis sets that assumed that the bound products are all linear molecules ($3n - 5$ vibrational modes). Figure 3 shows quite dramatically why the condensation of carbon is so facile and why it is biased toward addition of carbon, even in the presence of H and O. Here, the only molecule-dependent parameters are the vibrational frequencies themselves and the bond energies of C=C, C=O, and C–H bonds.

The complex lifetime, by itself, allows only a small percentage of C_3 to be formed by a direct association mechanism. However, under the high-

**Table II. Ab Initio Harmonic Vibrational Frequencies
for Linear Carbon Clusters**

Cluster	Harmonic Vibrational Frequency (cm^{-1})	Ref.
C_2	1940 (σ_g)	21
C_3	1367 (σ_g), 154 (π_u), 2311 (σ_u)	21
C_4	2345 (σ_g), 1022 (σ_g), 1740 (σ_u), 408 (π_g), 209 (π_u)	23
C_5	2220 (σ_g), 863 (σ_g), 2344 (σ_u), 1632 (σ_u), 222 (π_g), 648 (π_u), 112 (π_u)	21
C_6	2418 (σ_g), 1845 (σ_g), 721 (σ_g), 2190 (σ_u), 1327 (σ_u), 134 (π_g), 264 (π_g), 368 (π_u), 117 (π_u)	24
C_7	2376 (σ_g), 1745 (σ_g), 631 (σ_g), 2281 (σ_u), 2132 (σ_u), 1206 (σ_u), 598 (π_g), 157 (π_g), 710 (π_u), 240 (π_u), 73 (π_u)	21
C_9	2415 (σ_g), 2134 (σ_g), 1393 (σ_g), 496 (σ_g), 2338 (σ_u), 2084 (σ_u), 1803 (σ_u), 960 (σ_u), 658 (π_g), 252 (π_g), 114 (π_g), 783 (π_u), 567 (π_u), 187 (π_u), 49 (π_u)	21

NOTE: The frequencies for each cluster are listed in the order ν_1, ν_2, ...
ν_n. Very low-frequency (π) bending modes characterize this series of
clusters.

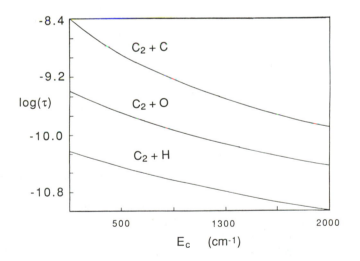

Figure 3. Collision complex lifetimes for the reactions of $[C_2 + X]$ (X = C, O, or H) as a function of energy above the C_2X dissociation threshold. 2000 cm^{-1} corresponds to a roughly 3000-K collision energy. The C_2 + C collision complex has a uniquely long-lived character. This calculation provides insight into the extremely facile nature of carbon condensation.

pressure conditions of the carbon arc, a much larger percentage of the C_3 will live long enough to experience a stabilizing collision with a helium or carbon atom. For reactions of the type C_n + C, with $n > 2$, the collision complex rarely requires a third body for stabilization, and condensation will quickly occur at or near the gas kinetic rate. Suzuki and co-workers (*19*) have pointed out a similar growth mechanism to describe the circumstellar chemistry of carbon cations.

Thus, the low-frequency degenerate bending modes in linear C_n clusters, coupled with high C_n binding energies, lead to extremely facile condensation kinetics, kinetics that are unique to the carbon system. As already stated, ab initio theory indicates that these early products of carbon condensation (<10 atoms) will have low-lying linear electronic states (*8, 20, 21*). Bonding in these clusters is predicted to be cumulenic (all bonds are C=C bonds), as opposed to acetylenic (alternating single and triple bonds). The even-numbered members of this series are calculated to exist as radicals, with $^3\Sigma$ ground states, and $^1\Sigma$ ground states are predicted for the odd-numbered chains. In addition, low-lying cyclic structures for the C_4, C_6, and C_8 clusters are also predicted (*22–25*). Much of this theory for the linear structures has been confirmed by high-resolution infrared spectroscopy (*13–15, 26–33*), and experimental data that have been applied to the determination of the structures of triplet linear C_4 and singlet linear C_9 are presented in Figures 4 and 5, respectively.

Molecular parameters derived from fits (*see* the section "Energy Level Expressions Used for Determination of Carbon-Cluster Molecular Parameters" for details) to this and similar spectra for other linear chains are presented in Table I. Analysis of the C_9 spectrum indicates that, although it is linear, the v = 1 level of the $\nu_6(\sigma_u)$ mode is heavily perturbed by a number of "dark" vibrational levels. This finding is consistent with the calculation that low-frequency bending modes lead to anomalously high vibrational state densities. For C_9, at an excitation energy of only 2000 cm^{-1}, equation 4 indicates that the vibrational state density is 10^8 cm^{-1}!

Step 2. From One Dimension to Two Dimensions

The next step in the condensation of carbon to produce the fullerenes is the formation of monocyclic rings, which are the calculated ground states for C_{10}–C_{20}. Early calculations predicted that linear structures, which are an inherently higher entropy morphology, would dominate this size range, at least up to n = 14, at the elevated temperatures necessary for graphite vaporization (*20*). However, several observations indicate that this prediction is incorrect. The first indication comes from a Δn = 4 intensity alternation in the mass spectrum of $C_n{}^+$ cations in this size range, which shows maxima at n = 11, 15, and 19 (*34*). This finding is easily explained by invoking aromaticity, as ring structures of these clusters are $4n + 2\pi$ electron systems. In addition, early structural calculations for these clusters indicated that the monocyclic rings (for $n < 14$) were only slightly lower in energy than the linear structures, and entropy contributions at high temperatures were thus quite important (*20, 35*). However, more sophisticated ab initio techniques now indicate that, for $n > 9$,

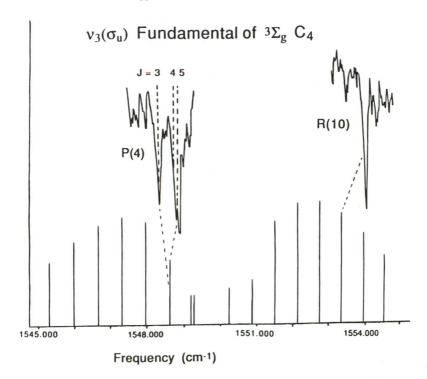

Figure 4. Fitted rovibrational stick spectrum and experimental data of the $\nu_3(\sigma_u)$ antisymmetric stretch fundamental of C$_4$ measured by IR diode laser absorption spectroscopy. Analysis of this spectrum indicates that C$_4$ has a low-lying (possibly ground) linear $^3\Sigma_g$ electronic state. Such linear triplet structures are predicted to dominate the C$_{2n}$ series for n < 5. (Reproduced with permission from reference 26. Copyright 1991 American Institute of Physics.)

the monocyclic rings are heavily favored (36). The temperature dependence of the equilibrium constant for the isomerization reaction

$$C_n \text{ (cyclic)} \xleftarrow{\quad} K_{eq} \xrightarrow{\quad} C_n \text{ (linear)}$$

may be calculated for a given cluster by using

$$K_{eq} = \frac{[q_{rot}(T)\,(\text{linear})]\,[q_{vib}(T)\,(\text{linear})]\,\exp\,(D_e/kT)}{[q_{rot}(T)\,(\text{cyclic})]\,[q_{vib}(T)\,(\text{cyclic})]\,\exp\,(D_e/kT)} \qquad (7)$$

Here, q_{rot} and q_{vib} are the rotational and vibrational partition functions, respectively; D_e is the calculated binding energy; k is the Boltzmann constant; T

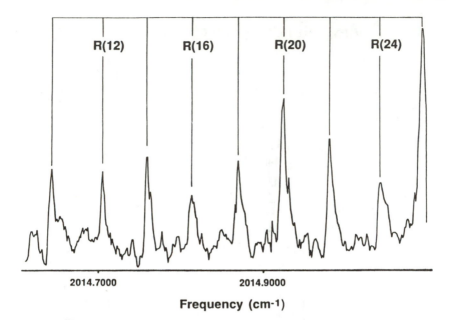

R(12)   R(16)   R(20)   R(24)

2014.7000   2014.9000

Frequency (cm⁻¹)

Figure 5. Experimental data of the $\nu_6(\sigma_u)$ antisymmetric stretch fundamental of linear $^1\Sigma_g$ C_9. The C_9 cluster is the largest cluster for which the ground state is predicted to be linear. (Reproduced with permission from reference 33. Copyright 1990 American Institute of Physics.)

is absolute temperature; and the electronic partition functions have been neglected (*37*). Structures and vibrational frequencies from experiment (*26*) (where available) and ab initio theory (*22–24, 36*) were used as input. The temperature dependence of K_{eq} is plotted in Figure 6 for both C_4 and C_{10}. For the C_4 cluster, the rhombic and linear structures are calculated to be almost isoenergetic, and the rhombic structure is slightly favored (by ∼0–5 kcal/mol) (*22–24*). Figure 6 shows that the high-temperature form is clearly the linear one, as expected. This calculation favored the rhombic structure by 2 kcal/mol. This result is consistent with the observation of triplet, linear C_4, produced in a high-temperature plasma by laser vaporization of graphite, by infrared diode laser absorption spectroscopy (Figure 4) (*26*). Recent predictions by Liang and Schaeffer (*36*) for the C_{10} cluster indicate that the energy difference between the monocyclic D_{5h} structure and the triplet chain is quite large (67 kcal/mol). Figure 6 shows the ring to be heavily favored past 3500 K. Thus, although entropy does play a significant role in determining the high-temperature structures of the smaller clusters, by $n = 10$, the higher entropy of the chain is not sufficient to overcome the higher binding energy of the monocyclic ring.

If monocyclic rings are the dominant intermediates in the region from C_{10} to C_{20}, then what chemical reactions lead into this size range? Once again,

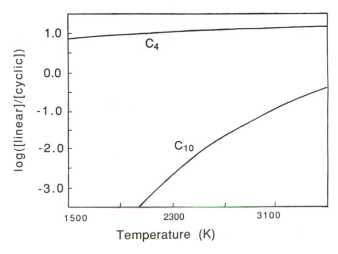

Figure 6. Plot showing the temperature dependence of the equilibrium ratios of a linear chain structure versus a monocyclic ring for the C_4 and C_{10} clusters. Even though cyclic C_4 is assumed to be slightly more stable than the linear structure (by ~2 kcal/mol) the linear structure dominates at the high temperatures present in the carbon arc. The cyclic structure of C_{10}, however, is predicted to be substantially more stable than the linear structure (by 67 kcal/mol) and continues to be dominant, even at 3500 K.

ab initio electronic structure calculations provide some strong clues to help answer these questions. Large amplitude motion in the degenerate bending (π_g and π_u) coordinates of the linear carbon cluster may lead to isomerization pathways from the linear structures to the monocyclic rings. For even-numbered clusters, however, such a mechanism seems unlikely, as it involves the spin-forbidden transition from a triplet to a singlet state. Recent calculations by Liang and Schaefer (*38*) on linear, even-numbered C_n ($n = 2$–10) indicate, however, that as n increases, the energy separation of the lowest singlet and triplet states becomes vanishingly small for C_8 and C_{10}. Andreoni et al. (*39*) used the Carr–Parrinello molecular dynamics approach to study the temperature-dependent behavior of the C_4 and C_{10} clusters. Upon heating rhombic C_4 to extreme temperatures (10,000 K) they found a linear–cyclic isomerization pathway that proceeds through a series of three-dimensional structures. C_{10}, on the other hand, was found to isomerize directly through a ring opening–closing mechanism at temperatures near 3000 K.

Some of the odd clusters appear to have such a pathway through their ground electronic states. This pathway is best understood by looking at the symmetries of the highest occupied molecular orbitals (HOMOs) of these clusters, shown in Figure 7. The symmetry of the HOMO alternates with respect to odd-numbered cluster size: C_3 and C_7 (and C_{11}) all have π_u HOMOs, but C_5 and C_9 have π_g HOMOs (*40*). Only the π_u HOMO provides the proper MO symmetry on the terminal carbons to facilitate isomerization to a monocyclic

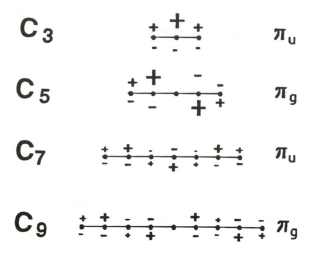

Figure 7. Symmetries of the highest occupied molecular orbitals (HOMOs) for odd-numbered linear C_n clusters. π_u and π_g alternate with respect to cluster size.

ring structure. In addition, the lowest frequency degenerate bending vibration for C_3 and C_7 is motion about the central carbon atom, and the force constants for this vibration are calculated (41) or observed (15) to be extremely small. Motion in the lowest frequency bending coordinates of C_3 has been measured by far-infrared spectroscopy (13) and by simulated emission pumping (15). These measurements indicate that the bending potential of C_3 is, indeed, extremely shallow and anharmonic.

One and two quanta of the lowest frequency (ν_{11}) bending mode of C_7 have been observed as hot bands associated with the strongly allowed $\nu_4(\sigma_u)$ antisymmetric stretch (Figure 8) (31). The ν_{11} coordinate is predicted to lie near 70 cm^{-1} (see Table II), and excitation of motion in this coordinate leads to extremely large amplitude (~45°) bending motion about the central carbon atom. Such large amplitude motion, at this low excitation energy, is apparently without precedent in strongly bound molecular systems. This result is a strong indication that a low-energy barrier isomerization pathway does exist for intraconversion between the linear chains and the monocyclic rings. Certainly a number of bimolecular processes may also lead to monocyclic rings, although such mechanisms have not been the subject of any investigations to date.

A Note on Puckered Rings

All structural calculations on the monocyclic rings indicate that they adopt planar, cumulenically bound morphologies. For C_6, Raghavachari et al. (25) calculated that a puckered structure, although more stable than either a linear $^3\Sigma$ form or a D_{6h} ring, is slightly less stable than their predicted D_{3h} monocyclic ground state. If such a puckered structure for any of the rings (with $n > 4$)

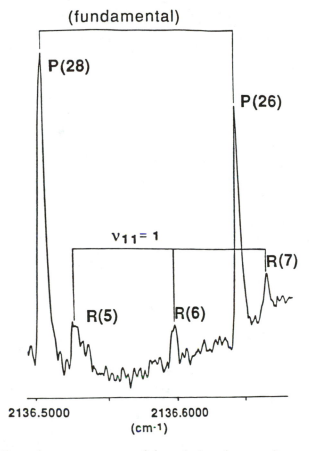

Figure 8. *Absorption measurement of the* $\nu_4(\sigma_u)$ *antisymmetric stretch and the associated* $\nu_{11}(\pi_u)$ *bending hot band of the* C$_7$ *cluster. For the pure stretching mode, only even rotational quanta are observed, and the P(26) and P(28) are shown. However, for the hot-band progression, even and odd rotational quanta are present, and J = 5–7 of the R branch are shown. Analysis of these (and two other) bands indicates that* C$_7$ *is extremely nonrigid. The low-frequency bends in this cluster may provide isomerization pathways for the formation of a monocyclic ring structure. (Reproduced with permission from reference 31. Copyright 1991 American Institute of Physics.)*

is a minimum on the potential energy surface, then the entropy arguments presented here for linear C$_n$ may be misleading. The entropy of puckered rings has been measured to be anomalously high, and this finding has been ascribed by Kilpatrick et al. (*42*) as being due to a rotation of the phase of the puckering about the circumference of the ring. Thus, the thermodynamics associated with pseudorotation may cause puckered monocyclic rings to dominate at high temperatures, even when the binding energies of the chains and rings are very close. Unfortunately, the low symmetry of a puckered ring makes very accurate ab initio treatments of such species quite formidable. However, such calculations would be extremely useful in sorting out the high-temperature structures of small carbon clusters.

Step 3. Growth from Two to Three Dimensions

Little is known about the clusters in the region from C_{21} to C_{31}. The fact that the carbon-arc synthesis works efficiently only in a limited pressure range is probably due to the chemistries of both this step and Step 4. The chemistries of Steps 1 and 2, on the other hand, appear to dominate the condensation of carbon in a wide variety of environments.

There do exist closed, fullerene-type solutions for all even-numbered clusters in this size range (with the possible exception of C_{22}). However, two properties characterize even the smallest fullerenes: (1) Even-numbered clusters are much more stable than odd-numbered clusters (2, 43), and (2) the primary photofragmentation pathway is loss of C_2 (9). None of the clusters in the C_{21}–C_{31} size range exhibit either property. Indeed, the primary photofragmentation pathway for these species is loss of C_3, which is also the primary photoproduct of the monocyclic rings and the linear chains (9, 44). However, the $\Delta n = 4$ intensity alternation, prevalent for the C_{10}–C_{20} size range, weakens throughout this region.

A model for the C_{21}–C_{31} clusters should draw from what is known about both the smaller clusters and the larger fullerenes, and it should reflect the chemical instability that appears to characterize these clusters. The C_{24} cluster shown in Figure 2, which represents this size region, is such a model. This cluster is composed of a 16-membered ring across which an eight-membered chain has added. Some cross-linking is beginning to occur, and already two five-membered rings and one six-membered ring have formed. The instability of this size range is reflected in that there is no way of satisfying all the carbon-atom valences in this structure, and hence it would be expected to take on carbon atoms and clusters quickly until a relatively stable closed fullerene structure is reached.

Step 4. Fullerene Growth

Experimental evidence for clusters with $n > 31$ indicates that their lowest energy structures are fullerene cages. There are, however, two possibilities for growth through this region: one that considers the fullerene structures as the chemical intermediates, and one that assumes higher energy, open structures as the reactive intermediates. In general, fullerene synthesis may be accomplished in a wide variety of environments, from laser ablation of graphite to sooting flames, and undoubtedly many different mechanisms contribute to their formation. However, the conditions under which high-yield synthesis of fullerenes is effected are quite limited. This fact indicates that one type of chemical process dominates the carbon arc, and that process must be either a closed-shell or an open-shell process.

The open-shell synthesis model, presented by Haufler et al. (4), argues that these clusters, at least up to C_{60}, are not fullerene cages, but are open graa)

b)

c)

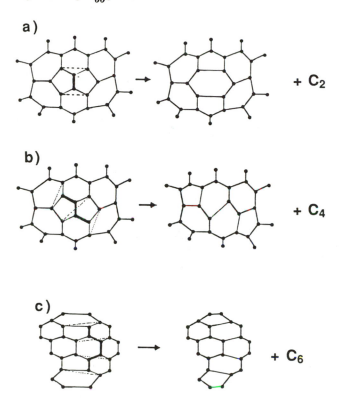

Figure 9. Possible mechanisms for fragmentation of spheroidal carbon shells. Shown are loss of (a) C_2, (b) C_4, and (c) C_6. Although these mechanisms do not require the fullerene cage to open, they are insufficient to describe closed-shell fullerene growth. Taken in reverse, they tend to maximize the number of adjacent pentagons. (Reproduced with permission from reference 9. Copyright 1988 American Institute of Physics.)

phitic sheets, similar to the $n = 21–31$ size range. The function of the buffer gas is to anneal these structures such that, as they grow, they are continually maximizing the number of pentagons while minimizing the number of adjacent pentagons. This model is appealing in that C_{60} and C_{70} are the first clusters that have no adjacent pentagons and no dangling bonds, and thus are "dead ends" in the cluster growth process.

The second possibility is that all clusters above a critical size (e.g., $n = 40$) are closed or nearly closed (for odd-numbered C_n) fullerene cages. Cluster growth proceeds via the addition of small carbon clusters without the cage ever opening. Here, the major driving force for chemical reaction is the minimization of adjacent pentagons (release of strain), and C_{60} and C_{70} are, again, the first stable products. O'Brien et al. (9) have proposed closed-shell mechanisms to explain the loss of C_2, C_4, C_6, etc., observed in fullerene photofragmentation (Figure 9). These mechanisms *cannot* be switched to explain fullerene growth

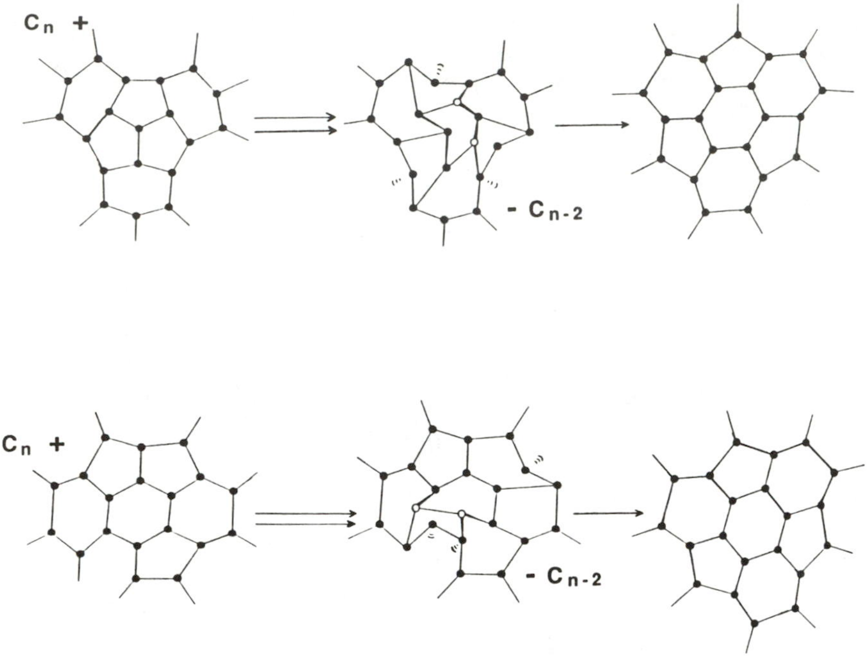

Figure 10. Possible mechanisms for growth of spheroidal carbon shells. At top is a mechanism by which the addition of a C_2 unit completely separates three adjacent pentagons. At bottom is a mechanism by which C_2 addition separates two adjacent pentagons. These mechanisms lead directly to C_{60} and C_{70} formation, as the end product in both schemes actually constitutes a fragment of either of the two fullerenes.

because, taken in reverse, they tend to maximize the number of adjacent pentagons.

A growth mechanism that tends to minimize the number of adjacent pentagons is shown in Figure 10. The top of Figure 10 shows how the addition of a C_2 unit may completely separate three adjacent pentagons and produce a structure that is actually a fragment of the C_{60} and C_{70} structures. This mechanism is applicable to all fullerenes smaller than C_{50}, all of which have three or more adjacent pentagons. The bottom of Figure 10 shows a similar mechanism, but in this case the addition of C_2 separates two adjacent pentagons, again producing a fragment of the C_{60} and C_{70} structures. This mechanism is applicable to all clusters up to C_{70}, excluding C_{60}.

These mechanisms require substantial reorganization of the molecular framework, and consequently, high activation barriers are expected. Thus, the high temperature of the carbon arc and the annealing properties of the buffer gas are once again critical to this synthesis. The laser vaporization synthesis of C_{60} is efficient (10% yield) only at very high (1000 K) temperatures (4). Such temperatures may be necessary to overcome the activation energy of these reactions. The appealing aspects of these mechanisms are

1. The lowest energy structures (fullerene cages) of these clusters are the reactants.
2. C_{60} and C_{70} are the first two clusters that do not have such a mechanism for growth.
3. This growth would be expected to occur in only a limited range of temperatures and pressures.
4. The structures produced naturally lead to C_{60} and C_{70}.

In addition, this mechanism explains why attempts at incorporating a metal atom into the internal cavity of a fullerene have been unsuccessful via the carbon-arc synthesis, because the structures that are of sufficient size to irreversibly trap a metal atom *are already closed.*

Energy Level Expressions Used for Determination of Carbon-Cluster Molecular Parameters

Many of the models discussed in this chapter rely on information derived from high-resolution infrared (IR) and far-infrared (FIR) spectroscopy of the unstable carbon-cluster intermediates involved in the carbon-condensation process. The following is a brief discussion of how molecular parameters are derived from the measured IR and FIR spectra.

Infrared spectroscopy of the linear carbon clusters is facilitated by the fact that C=C antisymmetric stretching fundamentals are typically very intense, with band origins in the region 1500–2300 cm^{-1}. The ^{12}C nucleus has spin $I = 0$. The Pauli exclusion principle thus allows only symmetric spin states for linear C_n and causes all rotational levels of a given vibronic state to have the same total parity. This condition implies that for the $^1\Sigma_g$ ground states of the linear odd C_n, only even J rotational levels are allowed. For the $^3\Sigma_g$ ground states of linear even C_n, only even N ($=J, J \pm S$) levels are allowed, where S ($= 1$) is the electron spin for a triplet molecule. In addition, the selection rule for rovibrational transitions is $\Delta J = \pm 1$. Thus, rovibronic bands for transitions from Σ_g ground states to $v = 1$ of σ_u vibrational levels are characterized by single (neglecting spin) P and R branches, with rotational lines separated by approximately $4B$, where B [$= h/(4\pi cI)$], in reciprocal centimeters, is the rotational constant of the cluster, c is the speed of light, and I is the moment of inertia about the axis of rotation. The σ_u ($v = 1 \leftarrow 0$) rovibrational transitions for linear odd C_n can be fit with the simple energy level expression:

$$E_{vr} = (v + \frac{1}{2})\nu_0 + B_\nu[J(J+1)] - D_\nu[J(J+1)]^2 + H_\nu[J(J+1)]^3 \quad (8)$$

Here v is the vibrational quantum number; ν_0 is the vibrational frequency; and B, D, and H are the rotational, quartic distortion, and sextic distortion constants, respectively. For even C_n, each corresponding rovibrational transition is actually split into three spin multiplets, corresponding to $N = J + 1, J$, and $J - 1$. The only even linear C_n for which molecular parameters have been determined, C_4, exhibits extreme Hund's case B coupling, meaning that the spin and rotational angular momentum are only weakly coupled. In practice, this weak coupling results in splittings that are are only resolvable for the lowest N transitions. Thus, equation 8, with N substituted for J, is sufficient for fitting all but the lowest N transitions for C_4. The observed splittings for small N are fit satisfactorily with the Schlapp expressions (45, 46):

$$E_{vrs} (J = N + 1) = E_{vr} - \left[\frac{2\lambda(N + 1)}{2N + 3} \right] + \gamma(N + 1) \qquad (9)$$

$$E_{vrs} (J = N) = E_{vr} \qquad (10)$$

$$E_{vrs} (J = N - 1) = E_{vr} - \left[\frac{2\lambda N}{2N - 1} \right] - \gamma N \qquad (11)$$

where λ and γ are spin–spin and spin–rotation constants, respectively, determined from electron spin resonance (ESR) spectroscopy of C_4 isolated in a rare gas matrix. For more accurate work, the Miller–Townes equations (47) should be used.

All linear C_n are predicted to have very low-frequency degenerate bending vibrations. Thus, even in a supersonically cooled cluster jet, hot bands arising from these bends, and associated with the strongly allowed σ_u vibrations, may be observable. In addition, direct transitions from the ground state to low-lying π_u bends are dipole-allowed, and may be investigated by FIR spectroscopy. The energy expression describing the rovibrational levels in a vibrationally excited π_u state is

$$E_{vr+(-)} = (v + \tfrac{1}{2})\nu_o + B[J(J + 1) - l^2] - D[J(J + 1) - l^2]^2 + H[J(J + 1) - l^2]^3$$
$$+(-) \tfrac{1}{2}\{q_l J(J + 1) + q_D[J(J + 1)]^2 + q_H[J(J + 1)]^3\} \qquad (12)$$

where q_l, q_D, and q_H are the l-type doubling constants associated with the rotational constant (B) and the sextic and quartic distortion constants (D and H), respectively. Here the $+(-)$ sign refers to states with even (odd) J.

Conclusions

The chemical model presented for the growth of small fullerenes is based upon the growth of closed fullerene structures, and leads directly to the formation of C_{60} and C_{70}. Much of this model is reinforced with both theoretical calculations and experimental observations; however, much of it remains untested. Although several of the structures of the unstable chemical intermediates discussed here have been unambiguously determined, many more have not. Certainly a detailed understanding of the structures and vibrational dynamics of the monocyclic ring structures will lead to new insights into their roles in the synthesis of the fullerenes. Accurate theoretical investigations into the nature of puckered monocyclic rings would, in particular, be extremely useful. In addition, the only reaction rates presented here were derived from spectroscopic information, and those calculations neglected anharmonic contributions to the vibrational frequencies. No actual kinetics for any of the reactions discussed here have been determined experimentally—even the $[C_2 + C \rightarrow C_3]$ reaction rate is unknown.

Finally, the chemical models for Steps 3 and 4 are largely untested. The model presented by Haufler et al. (*4*) for Step 4 and the model presented here for Step 4 have quite different implications for the synthesis of fullerenes that encapsulate metal atoms (*48*). Experiments that can differentiate between closed- and open-shell growth mechanisms for the high-yield synthesis would be particularly useful.

Acknowledgments

Much of the work presented here was generously supported by grants from the National Science Foundation, the Office of Naval Research, and the NASA Innovative Research Grants Program. I acknowledge the NASA SETI Foundation for my own support.

In addition, I thank R. J. Saykally for stimulating discussions, and, in particular, for pointing out the importance of puckered monocyclic rings. I also acknowledge R. E. Smalley for helpful discussions, especially concerning the open-shell growth mechanisms of Step 4. I acknowledge the assistance of J. Hawkins and A. Meyer in providing the apparatus for purification of the isotopically enriched material.

References

1. Krätschmer, W.; Fostiropoulos, K.; Huffman, D. R. *Chem. Phys. Lett.* **1990**, *170*, 167–171.

2. Kroto, H. W.; Heath, J. R.; O'Brien, S. C.; Curl, R. F.; Smalley, R. E. *Nature (London)* **1985**, *318*, 162–164.

3. Meijer, G.; Bethune, D. S. *J. Chem. Phys.* **1990**, *93*, 7800–7802.

4. Haufler, R. E.; Chai, Y.; Chibante, L. P. F.; Conceicao, J.; Jin, C.; Wang, L.-S.; Maruyama, S.; Smalley, R. E. *Mater. Res. Soc. Proc.* **1991**, in press.

5. Bevington, P. R. *Data Reduction and Error Analysis for the Physical Sciences*; McGraw-Hill: New York, 1959; pp 30–32.

6. Johnson, R. D.; Meijer, G.; Salem, J. R.; Bethune, D. S. *J. Am. Chem. Soc.* **1991**, *113*, 3619–3621.

7. Diederich, F.; Ettl, R.; Rubin, Y.; Whetten, R. *Science (Washington, D.C.),* **1991**, *252,* 548–551.

8. Weltner, W., Jr.; Van Zee, R. J. *Chem. Rev.* **1989**, *89,* 1713–1747.

9. O'Brien, S. C.; Heath, J. R.; Curl, R. F.; Smalley, R. E. *J. Chem. Phys.* **1988**, *88,* 220–230.

10. Gephardt, P.; Loffler, S.; Homann, K. *Chem. Phys. Lett.* **1987**, *137,* 306–310.

11. Creasy, W. R.; Brenna, J. T. *J. Chem. Phys.* **1990**, *92,* 2269–2279.

12. Marcus, R. A. *J. Chem. Phys.* **1952**, *20,* 359–365.

13. Schmuttenmaer, C. A.; Cohen, R. C.; Pugliano, N.; Heath, J. R.; Cooksy, A. L.; Busarow, K.; Saykally, R. J. *Science (Washington, D.C.)* **1990**, *249,* 897–900.

14. Kawaguchi, K.; Matsumura, K.; Kanamori, H.; Hirota, E. *J. Chem. Phys.* **1989**, *91,* 1953.

15. Smith, R. S.; Anselment, M.; Dimauro, L. F.; Frye, J. M.; Sears, T. J. *J. Chem. Phys.* **1988**, *89,* 2591.

16. Rohlfing, E. A. *J. Chem. Phys.* **1988**, *89,* 6103.

17. Robinson, P. J.; Holbrook, K. A. *Unimolecular Reactions*; Wiley-Interscience: London, 1972.

18. Marcus, R. A.; Rice, O. K. *J. Phys. Colloid Chem.* **1951**, *55,* 215–222.

19. Freed, K. F.; Oka, T.; Suzuki, H. *Astrophys. J.* **1982**, *263,* 718–722.

20. Pitzer, K. S.; Clementi, E. *J. Am. Chem. Soc.* **1959**, *81,* 4477–4485.

21. Raghavachari, K.; Binkley, J. S. *J. Chem. Phys.* **1987**, *87,* 2191–2197.

22. Magers, D. H.; Harrison, R. J.; Bartlett, R. J. *J. Chem. Phys.* **1986**, *84,* 3284–3290.

23. Ritchie, J. P.; King, H. F.; Young, W. S. *J. Chem. Phys.* **1986**, *85,* 5175–5182.

24. Martin, J. M. L.; Francois, J. P.; Gijbels, R. *J. Chem. Phys.* **1990**, *93,* 8850–8861.

25. Raghavachari, K.; Whiteside, R. A.; Pople, J. A. *J. Chem. Phys.* **1986**, *85,* 6623–6628.

26. Heath, J. R.; Saykally, R. J. *J. Chem. Phys.* **1991**, *94,* 3271–3273.

27. Bernath, P. F.; Hinkle, K. H.; Keady, J. J. *Science (Washington, D.C.)* **1989**, *244,* 562–564.

28. Heath, J. R.; Cooksy, A. L.; Gruebele, H. H. W.; Schmuttenmaer, C. A.; Saykally, R. J. *Science (Washington, D.C.)* **1989**, *244,* 565–567.

29. Moazzen-Ahmadi, N.; McKellar, A. R. W.; Amano, T. *J. Chem. Phys.* **1989,** *91,* 2140.

30. Heath, J. R.; Sheeks, R. A.; Cooksy, A. L.; Saykally, R. J. *Science (Washington, D.C.)* **1990,** *249,* 895–897.

31. Heath, J. R.; Saykally, R. J. *J. Chem. Phys.* **1991,** *94,* 1724–1729.

32. Heath, J. R.; Van Orden, A.; Kuo, E.; Saykally, R. J. *Chem. Phys. Lett.* **1991,** *182,* 17–20.

33. Heath, J. R.; Saykally, R. J. *J. Chem. Phys.* **1990,** *93,* 8392–8394.

34. Berkowitz, J.; Chupka, W. A. *J. Chem. Phys.* **1964,** *40,* 2735–2742.

35. Hoffman, R. *Tetrahedron,* **1966,** *22,* 521–538.

36. Liang, C.; Schaeffer, H. F. *J. Chem. Phys.* **1990,** *93,* 8844–8849.

37. McQuarrie, D. A. *Statistical Mechanics;* Harper Collins Publishers: New York, 1976.

38. Liang, C.; Schaefer, H. F., III. *Chem. Phys. Lett.* **1990,** *169,* 150–164.

39. Andreoni, W.; Scharf, D.; Giannozzi, P. *Chem. Phys. Lett.* **1990,** *173,* 449–454

40. Fan, Q.; Pfeiffer, G. V. *Chem. Phys. Lett.* **1989,** *162,* 472–476.

41. Brown, L. D.; Lipscomb, W. N. *J. Am. Chem. Soc.* **1977,** *99,* 3968–3979.

42. Kilpatrick, J. E.; Pitzer, K. S.; Spitzer, R. *J. Am. Chem. Soc.* **1947,** *69,* 2483.

43. Rohlfing, E. A.; Cox, D. M.; Kaldor, A. *J. Chem. Phys.* **1984,** *81,* 3322–3330.

44. Geusic, M. E.; McIlrath, T. J.; Jarrold, M. F.; Bloomfield, L. A.; Freeman, R. R.; Brown, W. L. *J. Chem. Phys.* **1986,** *84,* 2421–2422.

45. Herzberg, G. *Molecular Spectra and Molecular Structure II: Infrared and Raman Spectra of Polyatomic Molecules*; Van Nostrand Reinhold: New York, 1945.

46. Schlapp, R. *Phys. Rev.* **1937,** *51,* 342–347.

47. Miller, S. L.; Townes, C. H. *Phys. Rev.* **1953,** *90,* 537–545.

48. Heath, J. R.; O'Brien, S. C.; Zhang, Q.; Liu, Y.; Curl, R. F.; Kroto, H. W.; Tittel, F. K.; Smalley, R. E. *J. Am. Chem. Soc.* **1985,** *107,* 7779–7780.

Received August 12, 1991

Chapter 2

Crystalline Fullerenes

Round Pegs in Square Holes

R. M. Fleming, B. Hessen, T. Siegrist, A. R. Kortan, P. Marsh, R. Tycko, G. Dabbagh, and R. C. Haddon

AT&T Bell Laboratories, Murray Hill, NJ 07974

The fullerenes C_{60} and C_{70} act as spherical building blocks in crystalline solids to form a variety of crystal structures. In many cases, the icosahedral molecular symmetry of C_{60} appears to play little role in determining the crystal structure. In this chapter we discuss our results on the crystallography of pure and solvated fullerenes and some general features of fullerenes as building units in crystalline solids. For pure C_{60} or C_{70}, the face-centered cubic arrangement is preferred. In solvated crystals and compounds, the packing readily adapts to form non-close-packed structures.

The existence of the molecular carbon cluster C_{60} and higher molecular weight homologues was first deduced on the basis of spectroscopic measurements in the gas phase (*1*). These studies proposed that C_{60} was a hollow molecule with the shape of a truncated icosahedron, the familiar form of the seams on a soccer ball. Because of its hollow, high-symmetry structure, C_{60} was dubbed "buckminsterfullerene"; the name "fullerene" was used to refer to the class of hollow carbon spheroids. Studies of fullerenes have been recently accelerated by the dramatic discovery by Krätschmer et al. (*2*) that bulk quantities of C_{60} and lesser quantities of larger fullerenes such as C_{70} could be generated in a carbon arc struck in a partial pressure of helium. The availability of large quantities of material quickly led to additional structural studies based on ^{13}C NMR (*3–7*), IR (*8*), and Raman (*9*) spectroscopy as well as crystal structure determinations of compounds containing C_{60} (*10, 11*). A complete, atomic resolution crystal structure of pure C_{60} has not yet been accomplished, but the original soccer-ball model of C_{60} is now generally accepted.

The study of C_{60} and other fullerenes has proved to be exciting from a number of viewpoints. The structure, chemistry, and physical properties of C_{60} have all yielded surprising results and unique behavior. In this chapter we concentrate on the structural aspects of solid C_{60} and C_{70} and crystalline solvates of these molecules. We discuss the growth of "single" crystals of these materials and describe the types of disorder and twinning inherent in these structures. Because of the disorder, our structural studies are necessarily more focused on a description of the unit cells and the symmetry rather than a traditional crys-

0097–6156/92/0481–0025$06.00/0

tallographic structure with atomic resolution. In addition we also will describe the relationships between pure, crystalline C_{60} and the potassium intercalated compounds K_3C_{60} (12) and K_6C_{60} (13). The alkali intercalates have recently generated a great deal of interest in fullerenes because of the occurrence of conductivity and superconductivity in these materials (14–16).

Powder diffraction from solid C_{60} was obtained immediately after bulk synthesis and was used in the initial identification of the 10-Å d-spacing expected of a close-packed array of spherical molecules (2). Although the unit cell of solid C_{60} is now known to be face-centered cubic (fcc) (17), there was some early difficulty in assigning a unit cell and indexing the powder pattern. The confusion resulted from two curious features of the diffraction from C_{60}. The first was an absence of cubic (h00) reflections due to zeros in the form factor of a hollow molecule. The second was the presence of a shoulder on the cubic (111) reflection that has been attributed to planar defects in the structure (18). The shoulder coincidentally occurs at nearly the same d-spacing as a hexagonal close-packing (hcp) (100) reflection, a d-spacing that is not allowed in fcc.

Symmetry

The beautiful icosahedral symmetry of the C_{60} molecule with its alternating five- and six-membered rings is responsible for some of the most striking features of this new form of carbon. The point symmetry of the C_{60} molecule is $m\overline{3}5$ (I_h), which requires only one carbon position to describe the 60 sites contained in the molecule. As a consequence, the observation of a single narrow resonance line in solid-state ^{13}C NMR spectroscopy (3–7) on C_{60} at ambient temperature shows that the icosahedral point symmetry is correct and also shows that the C_{60} molecules undergo rapid, isotropic rotations at room temperature. The observation of molecular rotations in solid C_{60} raises questions about the expected symmetry in crystalline C_{60}. One could argue that, because of rapid rotation, one should treat the molecules in the unit cell as featureless spheres. On the other hand, C_{60} might be expected to be fully ordered and to adopt the highest symmetry allowed under the icosahedral point group. The fully ordered model could be consistent with the NMR observations if the rotations are steplike transitions between symmetry-equivalent positions with each molecule stationary for a finite time period.

First, assume that C_{60} is fully ordered with the maximum possible symmetry. Icosahedral symmetry does not allow translational periodicity, a fact that provides a primary motivation for the study of quasi-crystals (19). An icosahedral molecule can, however, form a crystallographic lattice by loosing symmetry elements. Figure 1 summarizes the crystalline, maximal subgroups of $m\overline{3}5$ (20). The highest order maximal subgroup of the icosahedron is $m\overline{3}$, a point group that retains 24 of the original 120 symmetry operations. This fact implies that in an ordered crystalline lattice, at least three carbon sites are needed to describe solid C_{60} instead of the one site needed to describe the isolated molecule. A common space group that occurs when local icosahedral symmetry is present, for example, in α-(AlMnSi) (21) and some virus molecules

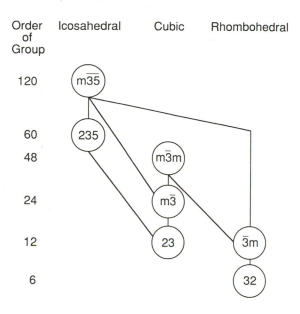

Figure 1. *Crystalline, maximal subgroups of the icosahedral point group* $m\overline{3}5$.

(22), is the body-centered cubic (bcc) structure $Im\overline{3}$. There is, however, no requirement that the symmetry be maximized, and exceptions such as boron, which crystallizes in the lower symmetry rhombohedral space group $R\overline{3}m$ (23, 24), are common. The bcc packing is plausible in that it allows planes of adjacent icosahedra to face each other, with holes at (¼ ¼ 0) and equivalent positions. For the truncated icosahedra, a bcc packing results in adjacent molecules having six-membered rings facing each other.

Contrary to these examples, X-ray diffraction shows that C_{60} crystallizes on an fcc rather than a bcc lattice (17), a result implying space group $Fm\overline{3}$ rather than $Im\overline{3}$. From the polyhedra-packing point of view, the occurrence of fcc over bcc is surprising, despite the fact that the density of fcc is somewhat higher. In fcc the sixfold rings of the truncated icosahedra no longer face each other: The sixfold rings are oriented along [111] directions and face the empty tetrahedral sites located at (¼ ¼ ¼). These sites contain potassium in K_3C_{60} (12). A drawing of a C_{60} molecule oriented so that the cubic [100] axis is normal to the page is shown in Figure 2. The three carbon positions labeled C1, C2, and C3 are the positions needed to describe the molecule in $m\overline{3}$. Even if the ordered description of C_{60} is correct, with molecules hopping between ordered orientations, the crystal will be prone to exhibit disorder. An example of the type of disorder one might expect is merohedral twinning (e.g., rotations by $\pi/2$ about [100]) of $Fm\overline{3}$. This merohedral twinning will increase the apparent symmetry of the material to $Fm\overline{3}m$. The merohedral twin operation is shown in Figure 2 by the dashed lines.

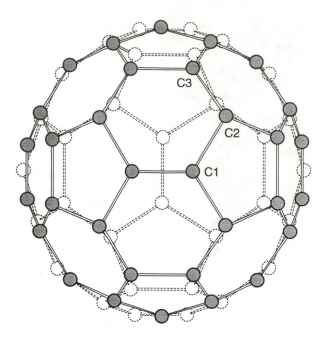

Figure 2. An idealized model of C_{60} oriented with the (h00) of the fully ordered model normal to the page. The dotted lines show the near registry of atoms that occurs upon a rotation of $\pi/2$ about (h00).

An alternative to the maximum symmetry model is to describe the C_{60} molecules as smooth, hard spheres. This description follows from the NMR results if we assume the rotations are continuous rather than steplike. The model also affords a rationalization for the selection of fcc over bcc by assuming that the structure selects the highest density available. Moreover, the symmetric shape of the C_{60} results in the absence of both a permanent dipole moment and local bond moments, a condition implying three-dimensional van der Waals bonding. A hard-sphere model does not, however, explain the selection of fcc over hexagonal close packing (hcp), a structure with an equivalent density. In other symmetric, noninteracting species such as the rare gases, the crystalline structures are universally fcc (except for He, which has quantum effects).

If one assumes a theoretical model for hard-sphere interactions such as the Lennard–Jones potential, hcp stacking is preferred over fcc, a conclusion that is contrary to observation (*25*). This contradiction shows that a Lennard–Jones model will not explain the molecular coordinations of either C_{60} or solid forms of noble gases. One must necessarily invoke higher order interactions to explain the packing of C_{60} as implicitly done in the calculations of Guo et al. (*26*), who calculated that the lowest energy structure of C_{60} occurs with an orientationally distorted fcc packing. As we will discuss, hard-sphere models are useful in determining the size of unit cells in solvated crystals, and they provide an adequate description of the room-temperature structure of C_{60}.

Sublimed Crystals

It is difficult to grow good crystals of pure C_{60} or C_{70} from solution because of the tendency of the material to form solvates or clathrates resulting from interactions with the solvent. The problems of solvation can be eliminated by using sublimation to grow crystals (*17*). When solid C_{60} is heated over a temperature gradient in an evacuated quartz tube (500 °C hot end), well-formed octahedral crystals sublime on the cool end of the tube. Powder X-ray diffraction from crushed crystals agrees with the early diffraction data (*2*). Precession X-ray photographs show full translational order and a face-centered cubic packing. Single-crystal diffractometer data are consistent with an apparent space-group symmetry $Fm\overline{3}m$ and a unit cell with a = 14.1981(9) Å (25 °C), Z = 4. As already discussed and illustrated in Figure 1, $m\overline{3}m$ point symmetry is too high to be compatible with icosahedral symmetry. Consequently, $Fm\overline{3}m$ apparent symmetry probably results from merohedral twinning and/or disorder of $Fm\overline{3}$, as illustrated in Figure 2.

A refinement based on single-crystal data and incorporating an equal and complete merohedral twinning operation proceeded as follows: Two carbon atoms put in at random positions were refined to form six-membered rings on the surface of a sphere with a 3.5-Å radius. A third carbon atom also refined to a position on the sphere around the origin, resulting in an approximate soccer-ball structure and R = 0.10, but with significant distortions, leading to unphysical C−C distances ranging from 1.2 to 1.7 Å. The refinement is not adequate to confirm the soccer-ball model, but it does confirm the structure of the fullerene as a hollow molecule with the carbon atoms distributed on the surface of a sphere with a 3.5-Å radius. The observed distortions are very likely to result from a combination of orientational disorder and the high thermal motion of the molecules in the solid at ambient temperatures, as indicated by NMR spectroscopy.

The lack of a well-ordered atomic structure of C_{60} at room temperature suggests that smooth hollow spheres might do equally well in refining the structure. This suggestion follows from the notion that the motion of C_{60} molecules at room temperature is isotropic. Our results as well as those of Heiney et al. (*27*) suggest that isotropic motion is the case. With hollow spheres and $Fm\overline{3}m$ symmetry, a slightly lower R-factor is obtained, about 8%. Practically speaking, little difference is seen in the magnitude of the errors using the ordered model and the smooth-sphere model. For calculations such as band structures, the choice of model largely depends on one's taste and the ease of application.

An interesting feature of the diffraction in pure C_{60} is the conspicuous absence of the (200) reflection, caused by the hollow nature of the molecule. For a spherically symmetric distribution of charge of radius r, the molecular form factor (f) is given by

$$f = \int 4\pi r^2 \, \sin\left|\frac{Qr}{Qr}\right| dr$$

Assuming a shell of charge where $\rho(r) = \delta(r - R)$, then the molecular form factor is given by $f \sim \sin(QR/QR)$ where Q is the X-ray momentum transfer 4π (sin $\theta)/\lambda$. Here θ is the scattering angle and λ is the X-ray wavelength. This function has zeroes at $Q = n\pi/R$, $n = 1, 2, \ldots$. Coincidentally, for $R = 3.55$ Å (the value obtained from single-crystal refinements), these zeros occurs almost exactly at the $(h00)$ reflections of C_{60}. If the lattice parameter is changed slightly so that the $(h00)$ reflections move off the zeros of the form factor, or if there is scattering from additional atoms or molecules in the unit cell other than C_{60}, the $(h00)$ reflections will be visible. This feature makes the intensity of the (200) reflection a very sensitive probe for various effects. For example, the application of hydrostatic pressure to crystalline fcc C_{60} changes the lattice parameters and causes the (200) reflection to appear (28).

The presence of extra electron density in the fcc unit cell (located either inside the sphere of the C_{60} molecule or in octahedral and/or tetrahedral holes between the close-packed layers) would also result in a nonzero intensity of the (200) reflection. Because the fcc intercalated compound K_3C_{60} has two components to the scattering factor, C_{60} and K, $(h00)$ peaks are observed (12). The refinement of the K_3C_{60} structure (9) gives the same C_{60} positions as the ordered twinned model discussed, a result suggesting that the molecules are spending finite amounts of time in symmetry positions.

In a number of situations, the motion of the C_{60} molecules is limited. One example is at low temperatures where powder X-ray diffraction and calorimetry experiments (27) show a phase transition at 249 K from fcc to primitive cubic in C_{60}. This feature appears to be associated with an orientational ordering of the C_{60} molecules in the cubic lattice. In the high-temperature fcc phase, all molecules are crystallographically identical, a characteristic causing all cubic reflections with $h + k + l =$ odd to be forbidden. The phase transition is characterized by the appearance of these forbidden reflections. This finding indicates that at low temperatures the molecules are no longer symmetry equivalent. ^{13}C NMR measurements show that the motional narrowing associated with tumbling of C_{60} molecules is not strongly affected at the phase transition, but the temperature dependence of the T_1 relaxation time shows a break (29). This result implies a change in the nature of the molecular motion at the phase transition, and the results are consistent with the model of dynamic hopping between symmetry positions in the low-temperature phase. Motion of the molecules continues until about 50 K, where the NMR line gradually broadens; such motion suggests that the transition from rotating molecules to fixed molecules occurs gradually and in an inhomogeneous manner.

A second situation in which the orientation of the C_{60} molecule appears to be fixed is in the intercalated compounds K_6C_{60} and Cs_6C_{60} (13). These compounds crystallize in a bcc unit cell, space group $Im\bar{3}$, the preferred crystallographic phase for packing icosahedra. The nearest-neighbor separation of

C_{60} molecules is nearly the same as in fcc, but there are 8 instead of 12 nearest neighbors. The body-centered structure allows an ordered structure with six-fold rings on adjacent molecules to face each other with four alkali atoms on each face of the cell. NMR measurements of this phase show a broadened line characteristic of a stationary C_{60} molecules.

An attempt to produce solvent-free C_{70} crystals by the same sublimation procedure described for C_{60} yielded small crystals with pentagonal faces, strongly reminiscent of pseudo-fivefold twins formed in electrodeposition of fcc Ag at high deposition rates (*30*). This result may suggest that single crystals of fcc C_{70} may be obtained by very slow sublimation. So far, the presence of extra diffraction lines in solvent-free powders of C_{60}–C_{70} mixtures, which can be indexed on an fcc cell with a larger lattice constant (a = 14.73 (2) Å, with a calculated nearest-neighbor distance of 10.415 Å), seems to support the possibility of C_{70} crystallizing on an fcc lattice. In addition, a powder obtained by subliming C_{70} yielded weak scattering that could be indexed by an fcc cubic cell. The C_{70} molecule is expected to be elongated by about 19% along one axis relative to the C_{60} molecule, and so the presence of a cubic (rather than tetragonal) lattice for C_{70} implies that the elongated molecule is disordered on the lattice. The observed lattice parameter is close to the value derived from the expected average dimension of the molecule.

Pentane Solvates

Slow diffusion of pentane vapor into benzene solutions of pure C_{60} or C_{70} produces well-formed prismatic crystals with an unusual 10-sided columnar habit (*31*), as shown in the scanning electron microscope (SEM) photograph of Figure 3. ^{1}H NMR spectroscopy of these crystals in CS_2 revealed the presence of a stoichiometric amount of n-pentane (with a <5 mol% impurity of benzene), an observation thus suggesting that these crystals can be formulated as fullerene · n-C_5H_{12} solvates. Precession X-ray photographs indicate full translational order and show a pseudo-10-fold diffraction symmetry in the plane normal to the columns. Diffraction in the plane containing the column axis shows ordering along the columns with a d-spacing of 10.08 (3) and 10.529 (6) Å for C_{60} and C_{70}, respectively. These correspond to the nearest-neighbor distances found in fcc C_{60} and C_{70} (the cubic [110] direction).

Although the 10-sided morphology and diffraction symmetry are reminiscent of features found in decagonal quasicrystalline materials (*32, 33*), the diffraction patterns can be completely indexed on the basis of a primitive monoclinic, non-close-packed unit cell together with a twinning operation. Any twinned close-packed structure is ruled out by the presence of a diffraction peak with a d-spacing of 15.8 Å for C_{60} (16.4 Å for C_{70}). In a close-packed structure with equivalent nearest-neighbor planes, the lowest order d-spacing [e.g., the cubic (111)] would be about 8 Å. A unit cell that is consistent with the diffraction data can be formed by shearing every second fcc (001) plane of the unsolvated cell in a [110] direction, as shown in Figure 4. This shearing

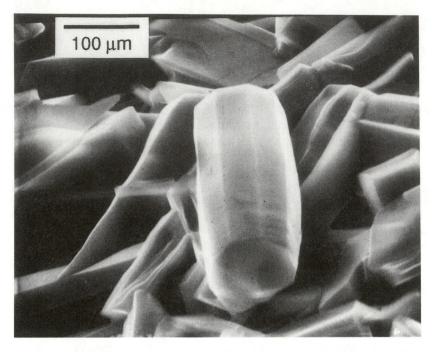

Figure 3. An SEM photograph of a pentane-solvated crystal of C$_{60}$. The wavy lines in the photograph are artifacts due to charging of the sample.

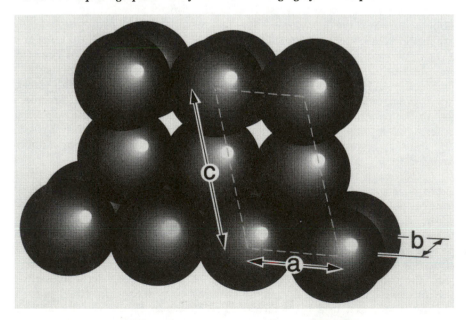

Figure 4. A unit cell that is consistent with the diffraction from pentane-solvated C$_{60}$ and C$_{70}$. The cell can be formed by shearing an fcc cell along a cubic [110] direction.

Table I. Lattice Parameters for Pentane-Solvated C$_{60}$ and C$_{70}$

Lattice	C$_{60}$ Primitive Monoclinic Obs.	Calc.	C$_{70}$ Primitive Monoclinic Obs.	Calc.	A-Centered Orthorhombic C$_{60}$ Obs.	C$_{70}$ Obs.
a	10.14 (3)	10.08	10.618 (6)	10.53	10.071	10.529
b	10.08 (3)	10.08	10.529 (6)	10.53	10.138	10.618
c	16.50 (5)	16.642	17.33 (1)	17.38	31.447	33.019
β	107.73 (3)	107.63	107.70 (3)	107.63	90	90

NOTE: Values of a, b, and c are given in angstroms, and β values are given in degrees. The numbers in parentheses are the standard error of the mean.

produces a primitive monoclinic cell with parameters of $a = b = D$, $c = 1.651D$, $\beta = 107.63°$ (with D the diameter of the fullerene contact sphere, e.g., 10.04 Å for C$_{60}$). (Further cell reduction of the model gives an A-centered orthorhombic cell, but we will continue to refer to the primitive monoclinic cell as it is more descriptive of the twinning operations.) The model parameters are remarkably close to the observed lattice parameters as obtained from twinned crystals on a four-circle diffractometer and summarized in Table I. The shearing of the fcc packing creates holes in the twinned structure that likely accommodate solvent. As there is one hole per two fullerene molecules in the unit cell, each hole should contain two pentane molecules to conform to the stoichiometry as determined by solution NMR spectroscopy.

The observed twinning and crystal morphology can be simulated by using the monoclinic unit cell just outlined. Twin domains can be obtained by first rotating the unit cell 180° around the a-axis, followed by a rotation around the b-axis of $180° - 2\beta* = 35.26°$, as illustrated in Figure 5. Successive pie-shaped domains obtained by this twinning operation can form a shape with nearly regular 10-fold morphology, as shown in Figure 6. The twinning rotates the cell by an irrational fraction, so the model predicts that twinned crystals should have a break or crack in their structure of $360° - 352.65° = 7.35°$. Measurements of the ω angles of reflections of different twin domains on a four-circle diffractometer indeed reveal such a deviation from perfect 10-fold symmetry, with an experimental gap-size of 5°.

The fact that the observed unit cell and twinning operation in these solvates can be simulated so well with a hard-sphere model indicates that the near 10-fold symmetry of the crystals is probably unrelated to the icosahedral molecular symmetry of C$_{60}$. It is also remarkable that C$_{70}$ crystallizes in exactly the same way as C$_{60}$, exhibiting an isotropic increase in the lattice parameters (4.7%, 4.5%, and 5.0% in a, b, and c, respectively). The observation of an isotropic increase of lattice parameters suggests that (at least at room temperature) the nonspherical C$_{70}$ molecule packs with random orientations, without alignment of the elongated axis of the C$_{70}$ molecule in the crystal. As discussed, the random nature of C$_{70}$ packing is also observed in unsolvated material.

The formation of solvates–clathrates by the fullerenes is by no means limited to pentane as included solvent. In various crystallization attempts

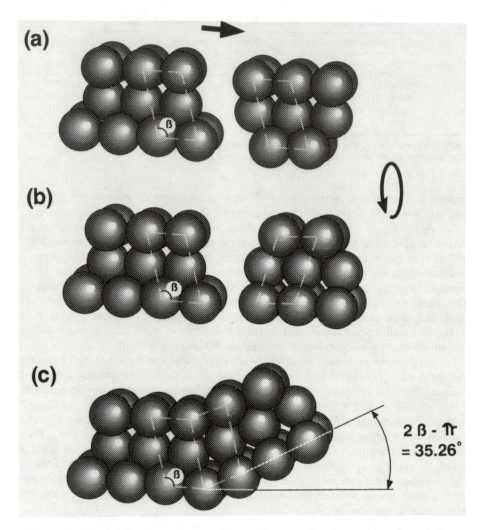

Figure 5. Model for the twinning observed in crystals of pentane-solvated C_{60} or C_{70}.

(including crystallizations from CS_2–pentane mixtures and even from pure benzene), we have observed crystals with a linear ordering of the fullerene molecules similar to that along the *a* axis in the pentane solvates. However, in most of these cases the diffraction patterns of the crystals in the perpendicular direction show strong azimuthal smearing of the reflections. This smearing may suggest that the monoclinic, sheared fcc structure is not unique to the pentane solvates, but that only in the pentane solvates does one have alignment of the tunnels to give full three-dimensional translational order.

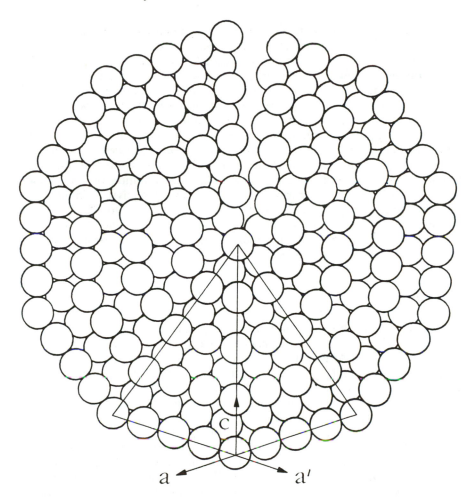

Figure 6. A collection of twins (obtained by the operation shown in Figure 5) illustrating the formation of a 10-sided crystal.

Crystallization of C_{60} from a CS_2–hexane mixture produced crystals that again indicate linear ordering of fullerene spheroids in one direction, but with full three-dimensional translational order and sixfold symmetry in the orthogonal direction. ^{1}H NMR measurements indicate that these are crystals of a stoichiometric hexane solvate $C_{60} \cdot C_6H_{14}$. The X-ray diffraction pattern can be indexed on a primitive hexagonal cell with lattice parameters $a = 33.651$ (4) Å and $c = 10.177$ (2) Å. These cell parameters are identical to those reported by Hawkins et al. (*34*) for crystals grown from pure hexane. Hawkins et al. assumed their crystal contained no solvent and attempted a structure determination with 13 C_{60} molecules in the cell on the basis of their density measurements. Our NMR data indicate that a better description may be obtained with

12 $C_{60} \cdot C_6H_{14}$ formula units per unit cell. In the diffraction data, the primitive cell is only weakly populated with six peaks at the cubic (111) d-spacing quite prominent. This weak population suggests that, like the case of pentane incorporation, the hexane solvate might be described by a modification of the fcc cell to accommodate hexane. However, unlike the pentane solvate, the twinning operations are true symmetry operations, resulting in a new unit cell.

Undoubtably other solvates of C_{60} and other fullerenes can be prepared. A solvate of C_{60} with cyclohexane has been reported (*35*) with a 24-Å fcc cell, but a complete structure of this solvate has not been described. Additionally, we have observed that pentane diffusion into CS_2 solutions of the fullerenes produces crystals of both C_{60} and C_{70} with the incorporation of CS_2 as indicated by SEM/EDAX (energy-dispersive X-ray analysis). Unlike pentane–benzene, the CS_2 solvates have different morphologies for C_{60} (flattened needles) and C_{70} crystals (stubby prisms). However, both of these crystal types suffer from extensive translational disorder and are prone to severe cracking, perhaps from loss of incorporated CS_2.

Summary

Despite early difficulties in synthesizing large quantities of fullerenes, these molecules have proved to be remarkably robust with surprising and unusual physical properties. At this stage, most work has focused on C_{60} because it is available in the largest quantities. C_{70} is less well studied, and higher molecular weight fullerenes have thus far received much less attention.

Both C_{60} and C_{70} crystallize in a variety of habits and form a number of solvates, salts, and compounds. Although one can make polycrystalline, unsolvated powder by carefully drying solutions of C_{60} or C_{70}, it is very difficult to grow unsolvated single crystals from solutions. Large "single" crystals of C_{60} can be obtained by sublimation. The crystals have a well-defined morphology but a large degree of orientational disorder, both static and dynamic. The disorder has made a traditional crystal structure of C_{60} at room temperature difficult. Both an ordered model where C_{60} molecules hop between symmetry positions and an orientationally disordered model where molecules are represented by spheres give similar refinement errors. True ordered structures of C_{60} have been obtained by stopping the molecular rotation by forming compounds (*10, 11*) or alkali intercalates (*12, 13*).

Well-formed, highly ordered, solvated crystals of C_{60} and C_{70} can be obtained by solution growth (*31, 34*). For crystals grown from pentane or hexane, solution NMR spectroscopy indicates that stoichiometric amounts of solvent are incorporated into the structure. For pentane, a twinned crystal with pseudo-10-fold symmetry can be indexed with a unit cell derived by shearing an fcc cell along [110]. The result is tunnels in the structure that can contain solvent molecules. Despite the 10-fold nature of the morphology and the diffraction, pentane-solvated fullerenes are not related to decagonal quasicrystals, but instead can be fully indexed with a primitive monoclinic, twinned cell. The structure of the hexane solvate is presently unsolved. The dimensions of the unit cell and the lack of dense diffraction spots suggest that the unit cell could be derived from a modification of the fcc cubic cell as in the pentane solvate.

Pure C_{70} and solvates of C_{70} crystallize on the same lattice as C_{60} under the same conditions at ambient temperatures. The elongated C_{70} molecules apparently have random orientations in the lattice so that C_{70} effectively packs as as spherical unit.

Acknowledgments

We acknowledge collaborations and discussions with S. J. Duclos, A. F. Hebard, M. L. Kaplan, S. H. Glarum, A. J. Muller, A. M. Mujsce, D. W. Murphy, K. Raghavachari, M. J. Rosseinsky, F. A. Thiel, and S. M. Zahurak.

References

1. Kroto, H. W.; Heath, J. R.; O'Brien, S. C.; Curl, R. F.; Smalley, R. E. *Nature (London)* **1985,** *318,* 162.

2. Krätschmer, W.; Lamb, L. D.; Fostiropoulos, K.; Huffman, D. R. *Nature (London)* **1990,** *347,* 354.

3. Tycko, R.; Haddon, R. C.; Dabbagh, G.; Glarum, S. H.; Douglas, D. C.; Mujsce, A. M. *J. Phys. Chem.* **1991,** *95,* 518.

4. Yannoni, C. S.; Johnson, R. D.; Meijer, G.; Bethune, D. S.; Salem, J. R. *J. Phys. Chem.* **1991,** *95,* 9.

5. Yannoni, C. S.; Bernier, P. P.; Bethune, D. S.; Meijer, G.; Salem, J. R. *J. Am. Chem. Soc.* **1991,** *113,* 3190.

6. Taylor, R.; Hare, J. P.; Abdul-Sada, A. K.; Kroto, H. W. *J. Chem. Soc. Chem. Commun.* **1991,** 1423.

7. Johnson, R. D.; Meijer, G.; Salem, J. R.; Bethune, D. S. *J. Am. Chem. Soc.* **1991,** *113,* 3619.

8. Krätschmer, W.; Fostiropoulos, K.; Huffman, D. R. *Chem. Phys. Lett.* **1990,** *170,* 167.

9. Bethune, D. S.; Meijer, G.; Tang, W. C.; Rosen, H. J. *Chem. Phys. Lett.* **1990,** *174,* 219.

10. Hawkins, J. M.; Meyer, A.; Lewis, T. A.; Loren, S. D.; Hollander, F. J. *Science (Washington, D.C.)* **1991,** *252,* 312.

11. Fagan, P. J.; Calabrese, J. C.; Malone, B. *Science (Washington, D.C.)* **1991,** *252,* 1160.

12. Stephens, P. W.; Mihaly, L.; Lee, P. L.; Whetten, R. L.; Huang, S.-M.; Kaner, R.; Diederich, F.; Holczer, K. *Nature (London)* **1991,** *351,* 632.

13. Zhou, O.; Fischer, J. E.; Coustel, N.; Kycia, S.; Zhu, Q.; McGhie, A. R.; Romanow, W. J.; McCauley, J. P., Jr.; Smith, A. B., III; Cox, D. E. *Nature (London)* **1991,** *351,* 462.

14. Haddon, R. C.; Hebard, A. F.; Rosseinsky, M. J.; Murphy, D. W.; Duclos, S. J.; Lyons, K. B.; Miller, B.; Rosamilia, J. M.; Fleming, R. M.; Kortan, A. R.; Glarum, S. H.; Makhija, A. V.; Muller, A. J.; Eick, R. H.; Zahurak, S. M.; Tycko, R.; Dabbagh, G.; Thiel, F. A. *Nature (London)* **1991,** *350,* 320.

15. Hebard, A. F.; Rosseinsky, M. J.; Haddon, R. C.; Murphy, D. W.; Glarum, S. H.; Palstra, T. T. M.; Ramirez, A. P.; Kortan, A. R. *Nature (London)* **1991,** *350,* 600.

16. Rosseinsky, M. J.; Ramirez, A. P.; Glarum, S. H.; Murphy, D. W.; Haddon, R. C.; Hebard, A. F.; Palstra, T. T. M.; Ramirez, A. P.; Zahurak, S. M.; Kortan, A. R.; Makhija, A. V. *Phys. Rev. Lett.* **1991,** *66,* 2830.

17. Fleming, R. M.; Hessen, B.; Kortan, A. R.; Siegrist, T.; Marsh, P.; Murphy, D. W.; Haddon, R. C.; Tycko, R.; Dabbagh, G.; Mujsce, A. M.; Kaplan, M. L.; Zahurak S. M. *Mater. Res. Soc. Symp. Proc.* **1991,** *206,* 691.

18. Luzzi, D. E.; Fischer, J. E.; Wang, X. Q.; Ricketts-Foot, D. A.; McGhie, A. R.; Romanow, W. J. *J. Mater. Res.,* submitted.

19. *See,* for example, *The Physics of Quasicrystals*; Steinhardt, P. J.; Ostlund, S., Eds.; World Scientific: Singapore, 1987.

20. *International Tables for Crystallography*, Hahn, T., Ed.; Reidel: Dordrecht, Netherlands, 1983; Vol. A, p 781.

21. Cooper, M.; Robinson, K. *Acta Cryst.* **1966,** *20,* 614.

22. *See,* for example, *Cold Spring Harbor Symposia on Quantitative Biology, Vol. XXXVI: Structure and Function of Proteins at the Three-Dimensional Level*; 1971, pp 433–503.

23. Decker, B. F.; Kasper, J. S. *Acta Cryst.* **1959,** *12,* 503.

24. Hoard, J. L.; Sullenger, D. B.; Kennard, C. H. L.; Huges, R. E. *J. Solid State Chem.* **1970,** *1,* 268.

25. Donohue, J. *The Structures of the Elements*; Krieger: Malabar, FL, 1982.

26. Guo, Y.; Karasawa, N.; Goddard, W., III *Nature (London)* **1991,** *351,* 464.

27. Heiney, P. A.; Fischer, J. E.; McGhie, A. R.; Romanow, W. J.; Denenstein, A. M.; McCauley, J. P., Jr.; Smith, A. B., III; Cox, D. E. *Phys. Rev. Lett.* **1991,** *66,* 2911.

28. Duclos, S. J.; Brister, K.; Haddon, R. C.; Kortan, A. R.; Thiel, F. A. *Nature (London)* **1991,** *351,* 380.

29. Tycko, R.; Dabbagh, G.; Fleming, R. M.; Haddon, R. C.; Makhija, A. V.; Zahurak, S. M. *Phys. Rev. Lett.* **1991,** *67,* 1886.

30. Pangarov, N. A. *Growth Cryst.* **1970,** *10,* 63.

31. Fleming, R. M.; Kortan, A. R.; Hessen, B.; Siegrist, T.; Thiel, F. A.; Marsh, P.; Haddon, R. C.; Tycko, R.; Dabbagh, G.; Kaplan M. L.; Mujsce, A. M. *Phys. Rev. B* **1991**, *44*, 888.

32. Bendersky, L. *Phys. Rev. Lett.* **1985**, *55*, 1461.

33. Kortan, A. R.; Thiel, F. A.; Chen, H. S.; Tsai, A. P.; Inoue, A.; Masumoto, M. *Phys. Rev. B* **1989**, *40*, 9397.

34. Hawkins, J. M.; Lewis, T. A.; Loren, S. D.; Meyer, A.; Heath, J. R.; Saykally, R. J.; Hollander, F. J. *J. Chem. Soc. Chem. Commun.* **1991**, 775.

35. Gorun, S. M.; Greaney, M. A.; Cox, D. M.; Sherwood, R.; Day, C. S.; Day, V. W.; Upton, R. M.; Briant, C. E. *Mater. Res. Soc. Symp. Proc.* **1991**, *206*, 659.

Received September 18, 1991

Chapter 3

Low-Resolution Single-Crystal X-ray Structure of Solvated Fullerenes and Spectroscopy and Electronic Structure of Their Monoanions

Sergiu M. Gorun[1], Mark A. Greaney[1], Victor W. Day[2], Cynthia S. Day[2], Roger M. Upton[3], and Clive E. Briant[3]

[1]Corporate Research Laboratories, Exxon Research and Engineering Company, Annandale, NJ 08801
[2]Crystalytics Company, P.O. Box 82286, Lincoln, NE 68501
[3]Chemical Design, Ltd., 7 West Way, Oxford OX2 0JB, United Kingdom

Single crystals of solvated C_{60} exhibiting long-range order were prepared and their X-ray structures were solved. Hollow, spheroidal, rigidly refined C_{60} molecules are shown to pack in an "expanded" cubic lattice that incorporates solvent in the octahedral and tetrahedral lattice voids. A less constrained refinement confirms two types of C–C bonds with lengths close to those determined by other methods. Strong near-IR bands of the coulometrically prepared C_{60}^- monoanion radical, observed at 917, 995, and 1064 nm, are assigned to the HOMO–LUMO transition. Raman vibrations of the C_{60}^- anion radical excited state are tentatively calculated to occur at 652 and 1507 cm^{-1}. C_{70}^- does not exhibit such bands, in agreement with theoretical calculations. Electron spin resonance spectra of C_{60}^- suggest that electron self-exchange occurs at room temperature.

The recent large-scale preparation of fullerenes (*1*) triggered an intense research effort for their characterization. To date, however, limited information is available on the structure of underivatized fullerenes. The electronic structure of fullerene anions (fullerides) is also of current interest, due, at least in part, to their reported superconductive properties (*2, 3*).

This chapter reports our preliminary data on the preparation and X-ray characterization of single-crystal phases of solvated fullerenes. We also report the bulk production of fulleride monoanion radicals; their spectroscopic properties are compared with those predicted by theoretical calculations.

0097–6156/92/0481–0041$06.00/0

Preparation, Spectroscopy, and Electronic Structure of C_{60}^- and C_{70}^- Anions

Pure fullerenes were prepared as previously described (4). Their spectroscopic properties matched those known in literature and, therefore, will not be reported again here. Previous reports (4–8) presented the electrochemical reduction of fullerenes. Using controlled potential coulometry, we are able to produce bulk quantities of fullerene ions. Our experimental setup consists of a PAR 173 potentiostat and model 179 digital coulometer. Pt mesh is used as the working electrode. Prior to electrolysis, methylene chloride–toluene solutions of C_n with or without Bu_4NPF_6 (supporting electrolyte) showed no electron spin resonance (ESR) signal. The potential was kept at −0.7 V vs. the saturated calomel electrode (SCE) (a value shown (4) by previous differential pulse polarography experiments to selectively produce C_{60}^- and C_{70}^-) until the current reached 90% of its limiting value. The bulk reduction process can be monitored by ESR and electronic spectroscopy. Anion solutions can be stored and transferred at room temperature under an inert atmosphere without decomposition.

The room-temperature ESR spectrum of C_{60}^- anion radical is shown in Figure 1a. A sharp signal on top of a broad one is observed, and this observation confirms the production of the radical anions. The depressed g factor of the sharp signal, $g = 1.9999$ (as compared to $g = 2.0023$ for the free electron), suggests spin-orbit coupling expected for triply degenerate HOMO–LUMO (highest occupied molecular orbital–lowest unoccupied molecular orbital) pairs (see Figure 2b). A Jahn–Teller effect that results in the reduction of the degree of orbital degeneracy may also favor spin-orbit coupling.

The C_{70}^- radical ion gives similar spectra, although somewhat more anisotropic, with $g = 1.9978$. We tentatively assign the sharp and the broad signal to the electron-exchanging free C_{60}^- anion and ion-paired $(Bu_4N)^+$ C_{60}^-, respectively. Electron self-exchange is likely to occur easily in these systems because theoretical calculations (9) predict little changes in C–C bond lengths of C_{60} upon addition of one electron. Thus, the Frank–Condon barrier for electron transfer should be minimal. Electron self-exchange and ion-pair formation resulting in similar spectra are well known for other hydrocarbon anions. For example, Figure 1b shows the ESR spectrum of the naphthalene–naphthalenide anion (potassium salt) radical pair (10).

Further probing into the electronic structure of the fullerene ions was achieved by comparing the electronic spectra of C_{60}^- and C_{70}^- anion radicals to those of the parent molecules. The spectra of C_{60} and C_{60}^- anion radical (Figure 2a) clearly show strong near-IR bands for the anion radical (molar extinction coefficient, in reciprocal mole-liters, in parenthesis) at 917 (4.6×10^3), 995 sh (5.1×10^3), and 1064 (12×10^3) nm. These bands disappear upon electrochemical reoxidation of C_{60}^- to C_{60} and appear again upon reduction. This process can be repeated numerous times. No such bands are observed for

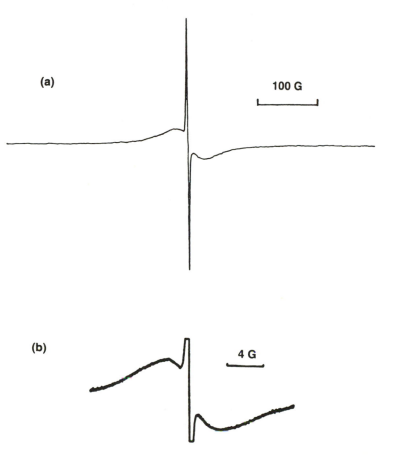

(a)

100 G

(b)

4 G

Figure 1. a: Room-temperature ESR spectrum of C_{60}^- monoanion radical in CH_2Cl_2–toluene. b: ESR spectrum of the naphthalene monoanion radical. (Part a reproduced from reference 19. Part b reproduced from reference 10. Copyright 1991 American Chemical Society.)

C_{70}^-. These observations can be explained by examining the C_{60} and C_{70} orbital splittings of Figure 2b (11–13). Assuming, in agreement, with theoretical calculations, that the I_h symmetry of C_{60} is not changed upon reduction, then the t_{1u} and t_{1g} orbitals of C_{60} can be used, to the first approximation to describe C_{60}^-, too. Thus, addition of one electron to t_{1u} makes the t_{1u}–t_{1g} transition possible and electric dipole allowed. If we ignore a (likely small) Jahn–Teller distortion, we can use the C_{60} orbital energies to calculate electronic transitions of C_{60}^-.

The wavelengths of the calculated t_{1u}–t_{1g} transition vary as a function of computational methods and molecular geometry from approximately 1200 to 2100 nm (11, 12, 14–16). The strongest observed near-IR band at 1064 nm is not far from this range, especially considering the computational uncertainties and our approximations. Thus, this strong band could be tentatively assigned

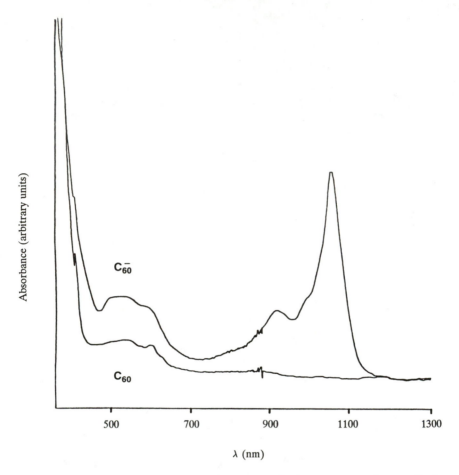

Figure 2a: Near-IR spectrum of C_{60} and C_{60}^- monoanion radical. (Reproduced from reference 19. Copyright 1991 American Chemical Society.)

to the allowed (0–0) t_{1u}–t_{1g} transition. The absorptions at 917 and 995 nm might be due to a combination of two phenomena: (1) a Jahn–Teller mechanism lifts the degeneracy of the triply degenerate t orbitals, and (2) vibrational levels within the excited state of C_{60}^- radical anion are well separated. In either case, the orbital or vibrational level splittings are 0.187 and 0.081 eV. These values correspond to 1507 and 652 cm^{-1}, respectively.

However, if we assume that vibronic coupling occurs, group theory helps in the interpretation of the near-IR spectrum of C_{60}^-. Assuming again t_{1u} and t_{1g} symmetry for C_{60}^- radical anion HOMO and LUMO, respectively, and using the symmetries of the fundamental vibrations of I_h molecules, it can be shown that transitions between vibronically coupled electronic–IR levels are symmetry-forbidden, but their Raman counterparts are allowed. Experimentally (*17*), Raman bands are observed *inter alia* at 1470, 1575, 496, and 710

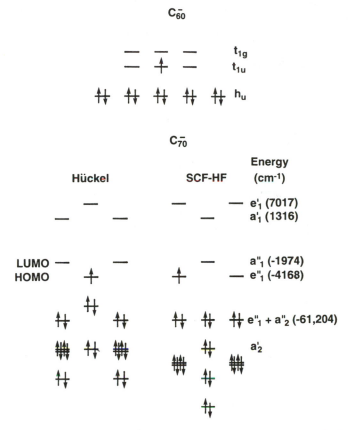

Figure 2b: HOMO−LUMO orbitals for C_{60}^- and C_{70}^- radical anions assuming no Jahn−Teller effect. For C_{70} the Hückel and SCF−HF results are shown. Energies, in reciprocal centimeters, are given only for the SCF−HF case. (Adapted from references 11−13).

cm^{-1}. Using these values and setting the 1064-nm near-IR band as the reference point, we calculate that $(0-n)$ vibronic transitions should be observed at 920, 911, 1011, and 989 nm. The first two calculated wavelengths are within experimental error of the observed 917-nm band; the last two are close to the observed 995-nm band. We can, therefore, tentatively assign the 917- and 995-bands to $(0-n)$ $t_{1u}-t_{1g}$ transitions.

Because we assumed little geometrical change in C_{60}^- vs. C_{60}, the observation of Raman bands for excited C_{60}^- radical anion close to those found in C_{60} is to be expected. This self-consistency does not prove the tentative assignments, but the agreement may not be fortuitous. Indeed, irradiation of matrix-trapped fullerene at 77 K also results in the production of anions

(18). The reported near-IR spectra independently seem to confirm our assignments, but the spectral interpretation is complicated by the potential presence of anions with different charges, trapped electrons, and dimer radical anions. Further details are published elsewhere *(19)*.

Using the self-consistent field, Hartree–Fock (SCF–HF) derived orbital energies of Figure 2b, we calculate the HOMO–LUMO transition for C_{70}^- radical anion to occur in the IR at approximately 2200 cm^{-1}. Preliminary experiments failed to detect new IR bands in solutions of C_{70}^- radical anion. According to group theory, this transition should be strongly plane (*xy*) polarized if C_{70}^- retains the C_{70} symmetry.

X-ray Studies on the Structure of Solvated C_{60}

Cubic Data-Set Refinement. We previously reported *(4)* the low-resolution single-crystal X-ray crystal structure of the first fullerene solvate, C_{60}–C_{70}·cyclohexane. Single crystals of cyclohexane solvates obtained from purified C_{60} extracts were also reported to be quasi-isomorphous with those containing C_{60}–C_{70} mixtures: a = 28.144 (15) vs. 28.216 (9) Å in the cubic space group *Fm3m* at 25 °C. The crystals were of reasonably good quality and exhibited long-range order.

The structure of solvated pure C_{60} is now solved and refined in the cubic system by using computer modeling and, initially, theoretically obtained C_{60} coordinates (Newton, M. D., Brookhaven National Laboratories, personal communication). C_{60} molecules, treated as rigid bodies, were placed at various sites in the cubic lattice, and refinement was attempted. The only convergent solution was obtained when the fullerenes were placed at [0, 0, 0] and [0.500, 0, 0]. Upon refinement, the positional parameters of the C_{60} placed at [0.500, 0, 0] (the origin defining C_{60} was fixed) shifted slightly to [0.504 (4), 0.038 (3), 0.005 (2)] and thus confirmed the solution and modeling studies. Common variable isotropic thermal parameters assigned to atoms belonging to the same C_{60} molecule refined to 3 (1) and 9 (1) $Å^2$ for the C_{60} at [0, 0, 0] and [0.504 (4), 0.038 (3), 0.005 (2)], respectively. Interestingly, additional observed disordered carbon peaks in difference Fourier maps define spherical electron-density-free voids around the [0.25, 0.25, 0.25] and [0.75, 0.25, 0.25] positions with approximately 6-Å diameters, in accord with the modeling studies that indicated the possibility of C_{60} at these positions. The final refinement of 14 variables (one scale factor, three rotational and one thermal parameter for each of the two independent C_{60}s; three translational parameters for the nonorigin C_{60}; one common occupancy factor; and one thermal parameter for the additional disordered atoms) using 106 observed unique reflections with intensities $I > 3\sigma(I)$ ($\sigma(I)$ is the estimated standard deviation of I) resulted in discrepancy factors R_1 and R_2 of 0.109 and 0.093, respectively. Further details of the refinement are published elsewhere *(20)*. A good model for the disordered cyclohexane molecules could not be found. Partial occupancy by C_{60} of the [0.25, 0.25, 0.25] and [0.75, 0.25, 0.25] sites resulted in the formation of channels along the sides of the unit cell. These channels, having a 7.1-Å diameter, can accommodate cyclohexane molecules, which have approximately 7-Å van der Waals diameter for spherical disorder.

Figure 3 shows the crystal-packing environment of C_{60}. The small spheres represent the sites for the cyclohexane molecules. Full occupancy by C_{60} of all available sites led to the formation of octahedral and tetrahedral voids. Figures 4a and 4b show the fit of a single cyclohexane in these spaces, respectively. Graphical manipulations of the solvent in the octahedral voids resulted in a preferential orientation for the cyclohexane, but this model could not be confirmed crystallographically, perhaps because the energy differences are small compared to kT.

The closest C_{60} nonbonding contacts, 12.218 Å, are significantly longer than the approximately 10.7-Å van der Waals radius of C_{60}. These larger separations seem to be solvent-induced and quasi-independent of the size of the fullerenes. Indeed, inclusion of a C_{70} molecule in the model based on the theoretically obtained atomic coordinates (Raghavachari, K., AT&T Bell Laboratories, personal communication) (*see* Figure 5) does not disrupt the C_{60} network, as evidenced by the isomorphism observed previously for the C_{60}–C_{70} crystals.

Interestingly, this work demonstrates that the solvent-free C_{60} cubic structure can be "expanded" in a solvated cubic structure by incorporation of spherically disordered solvent molecules. This expansion shifts the C_{60} fullerenes apart isotropically. The critical temperature (T_c) of the recently discovered C_{60}-based superconductors might be a function of the C_{60}–C_{60} distances in the crystalline lattice (*2, 3*), and this distance should be regulated by the size of other ions present in the tetrahedral and octahedral spaces of the cubic lattice. Thus, the discovery that expanded structures form single crystals exhibiting reasonable long-range order might be exploited in the design of new materials.

C–C Bond Length Refinement in C_{60}.

Because C_{60} moieties were successfully refined as rigid bodies in cyclohexane solvates of pure C_{60} and mixed C_{60}–C_{70}, it was decided to attempt refinement for the two types of C–C bonds believed to be present in nonderivatized C_{60} (*21*). The data set for solvated C_{60} crystals was chosen for this study because C_{70} molecules were not present and the disordering possibilities in the lattice were therefore fewer. The 106 independent cubic reflections for solvated C_{60} with $I > 3\sigma(I)$ were expanded to a triclinic set of 1679 reflections with $I > 3\sigma(I)$. This data set was used to refine two independent C_{60} units (one at [0, 0, 0] and the other at [0.500, 0, 0]) as rigid bodies and to locate 278 partial-occupancy disordered carbon atoms with procedures similar to those described previously with the 106 unique cubic reflections. Program limitations of 400 independent atoms prevented the inclusion of more disordered atoms.

During this refinement, the center of gravity of the first C_{60} was fixed at the origin, and that of the second C_{60} was refined to [0.513, 0.044, −0.008]. Starting with this model (two idealized rigid-body C_{60} molecules and 278 fixed partial-occupancy carbon atoms), the positions for carbon atoms of the first C_{60} molecule were then allowed to vary in cycles of constrained least-squares refinement in which the hexagon–hexagon (B–C in Figure 6) and hexagon–pentagon (A–B in Figure 6) C–C bonds were tied to two independent free variables. The second C_{60} molecule was refined as a rigid body with idealized geometry. Next, the positions for carbon atoms of the second C_{60} molecule were allowed to vary, and its two types of C–C bonds were again tied to two independent free variables. The first C_{60} was refined as an idealized rigid body. This alternate refinement was repeated until self-consistency was achieved, that is, the bond-length variations were less than 1 estimated standard deviation. For both C_{60} units, the hexagon–hexagon connecting C–C bonds refined to smaller values than the hexagon–pentagon connecting C–C bonds (1.43 (2) vs. 1.53 (1) Å and 1.45 (2) vs. 1.50 (2) Å for the first and second C_{60} units, respectively). The average values for these bonds are shown in Table I and compared to values determined via other techniques.

The hexagon–hexagon and hexagon–pentagon connecting bonds for the two independently refined C_{60}s show good internal agreement. Their average, 1.44 (2) and 1.51 (2) Å, are only slightly longer than their counterparts that result from theoretical calculations, NMR, and EXAFS (extended X-ray absorption fine structure) measurements. They are, however, longer than their counterparts in osmylated and platinated C_{60}. Thus, because different C–C bond lengths were resolved for both independent C_{60}s, and their relative lengths agree with other literature data, the presence of lattice solvent may decrease the rotational disorder of C_{60} to the point that attempts to model it might be successful.

Our refinement results are reasonably consistent with data obtained via other techniques on underivatized C_{60} and single-crystal X-ray results of derivatized (metalated) C_{60}. Solvated C_{60} crystals were initially claimed to lack long-range order (28) but, very recently, another aliphatic solvent, pentane, was shown to yield ordered C_{60} solvates (29). Furthermore, the room-temperature refinement of an unsolvated C_{60} structure yielded "unphysical bond distances" (30), and solid-state structures of solvent-free C_{60} were found to vary as a function of conditions of crystallization (31). The aforementioned crystallographic and modeling results combined with the observation of an unusually high (249 K) ordering transition temperature for pure C_{60} (32) lead us to postulate that even weak solvent–solute (C_{60}) interactions might be responsible for increased room-temperature C_{60} ordering. Nonplanar, nonaromatic solvents seem to favor higher degrees of crystallinity. The absolute value of our C–C bonds may not be accurate, but higher accuracy may result upon inclusion of all the atoms in the refinement and/or collecting X-ray data at lower temperatures.

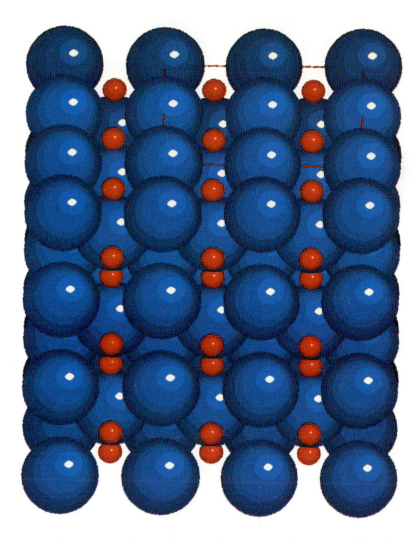

Figure 3. Modeling studies of the crystal-packing environment of C_{60} in the cubic lattice. Large spheres: C_{60}; small spheres: center of gravity of cyclohexane molecules.

a

b

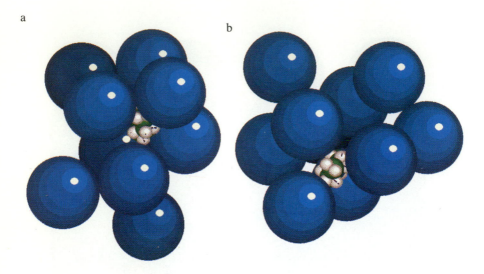

Figure 4. Modeling of cyclohexane fit in the (a) octahedral and (b) tetrahedral voids of C_{60} cubic lattice.

Figure 5. Model of C_{70} fitted in the C_{60} cubic lattice.

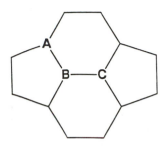

Figure 6. The unique portion of the C_{60} skeleton showing the labeling scheme used in Table I.

Table I. Bond Lengths (Angstroms) in C_{60} Compounds

Compound	AB^a	BC^a	Method	Ref.
C_{60}–cyclohexane	1.51 (2)	1.44 (2)	X-ray	this work
C_{60}	1.45 (2)	1.40 (2)	NMR	22
C_{60}	1.42 (1)	1.42 (1)	EXAFS	27
C_{60}–osmylated	1.432 (5)	1.388 (9)	X-ray	23
C_{60}–Pt derivative	1.45 (3)	1.39 (3)	X-ray	24
C_{60}	1.36–1.45	1.34–1.38	theor. calc.	25, 26

NOTE: The numbers in parentheses are the estimated standard deviations of the last digit.
[a]Bonds are shown in Figure 6.

Conclusions

In conclusion, single crystals of C_{60}–C_{70} and C_{60}^- solvates exhibiting long-range order allowed us to demonstrate the hollow, convex shape of C_{60} via X-ray techniques. Modeling the rotational disorder has allowed refinement of two different C–C bond lengths for two independent C_{60} molecules in C_{70} free crystals, thus confirming the truncated icosahedral arrangement of the carbon atoms. Both C_{60}^- and C_{70}^- radical monoanions have been prepared in bulk quantities by coulometry. ESR spectra of C_{60}^- suggest that facile electron self-exchange occurs at room temperature. Electronic spectroscopy of the anions reveal strong HOMO–LUMO transitions for C_{60}^- but not for C_{70}^- in the near-IR region. For the C_{60}^- radical anion, Raman vibration frequencies might be derived from the electronic near-IR spectrum. These results are in agreement with and confirm theoretical calculations.

Acknowledgments

We thank D. M. Cox, K. Creegan, R. D. Sherwood, and P. A. Tindall for a generous gift of fullerenes. M. Newton and K. Raghavachari are acknowledged for providing the calculated coordinates of C_{60} and C_{70}, respectively. G. George, A. Kaldor, and H. Thomann are thanked for helpful discussions.

References

1. Krätschmer, W.; Lamb, L. D.; Fostiropoulos, K.; Huffman, D. R. *Nature (London)* **1990**, *347,* 354.

2. Hebard, A. F.; Rosseinsky, M. J.; Haddon, R. C.; Murphy, D. W.; Glarun, S. H.; Palstra, T. T. M.; Ramirez, A. P.; Kortan, A. R. *Nature (London)* **1991,** *350,* 600.

3. Stephens, P. W.; Mihaly, L.; Lee, P. L.; Whetten, R. L.; Huang, S-M.; Kaner, R.; Diederich, F.; Holczer, K. *Nature (London)* **1991,** *351,* 632.

4. Gorun, S. M.; Greaney, M. A.; Cox, D. M.; Sherwood, R.; Day, C. S.; Day, V. W.; Upton, R. M.; Briant, C. E. *Proc. Mater. Res. Soc. Boston* **1991,** *206,* 659.

5. Haufler, R. E.; Conceicao, J.; Chibante, L. P. F.; Chai, Y.; Byrne, N. E.; Flanagan, S.; Haley, M. M.; O'Brien, S. C.; Pan, C.; Xiao, Z.; Billups, W. E.; Ciufolini, M. A.; Hauge, R. H.; Margrave, J. L.; Wilson, L. J.; Curl, R. F.; Smalley, R. E. *J. Phys. Chem.* **1990,** *94,* 8634.

6. Wudl, F.; Allemand, P. M.; Koch, A.; Rubin, F.; Diederich, F.; Alvarez, M. M.; Anz, S. J.; Whetten, R. L. *Proc. Mater. Res. Soc. Boston,* **1990,** abstract.

7. Ibid. *J. Am. Chem. Soc.* **1991,** *113,* 1050.

8. Allemand, P. M.; Srdanov, G.; Koch, A.; Khemani, K.; Wudl, F.; Rubin, Y.; Diederich, F.; Bethune, D. S.; Meijer, G.; Tong, W. C.; Rosen, H. J. *Chem. Phys. Lett.* **1990,** *174,* 219.

9. Negri, F.; Orlandi, G.; Zerbetto, F. *Chem. Phys. Lett.* **1989,** *144,* 31.

10. Chang, R.; Johnson, C., Jr. *J. Am. Chem. Soc.* **1986,** *88,* 2338.

11. Larsson, S.; Volosov, A.; Rosèn, A. *Chem. Phys. Lett.* **1987,** *137,* 501.

12. Haddon, R. C.; Brus, L. E.; Raghavachari, K. *Chem. Phys. Lett.* **1986,** *125,* 459.

13. Scuseria, G. E. *Chem. Phys. Lett.* **1991,** *180,* 446.

14. Satpathy, S. *Chem. Phys. Lett.* **1986,** *130,* 545.

15. Rosèn, A.; Wästberg, B. *J. Chem. Phys.* **1989,** *90,* 2525.

17. Bethune, D. S.; Meijer, G.; Tang, W. C.; Rosen, H. J.; Golden, W. G.; Seki, H.; Brown, C. A.; de Vries, M. S. *Chem. Phys. Lett.* **1991,** *179,* 181.

18. Kato, T.; Kodama, T.; Shida, T. *Chem. Phys. Lett.* **1991**, *180*, 446.

19. Greaney, M. A.; Gorun, S. M. *J. Phys. Chem.* **1991**, *95*, 7142.

20. Gorun, S. M.; Creegan, K. M.; Sherwood, R. D.; Cox, D. M.; Day, V. W.; Day, C. S.; Upton, R. M.; Briant, C. E. *J. Chem. Soc. Chem. Commun.* **1991**, in press.

21. Kroto, H. W.; Heath, J. R.; O'Brien, S. C.; Curl, R. F.; Smalley, R. E. *Nature (London)* **1985**, *318*, 162.

22. Yannoni, C. S.; Bernier, P. P.; Bethune, D. S.; Meijer, G.; Salem, J. R. *J. Am. Chem. Soc.* **1991**, *113*, 3190.

23. Hawkins, J. M.; Meyer, A.; Lewis, T. A.; Loren, S.; Hollander, F. J. *Science (Washington, D.C.)* **1991**, *252*, 312.

24. Fagan, P. J.; Calabrese, J. C.; Malone, B. *Science (Washington, D.C.)* **1991**, *252*, 1160.

25. Weltner, W., Jr.; van Zee, R. J. *Chem. Rev.* **1989**, *89*, 1713.

26. Scuseria, G. E. *Chem. Phys. Lett.* **1991**, *176*, 423.

27. Cox, D. M.; Behal, K.; Creegan, K.; Disko, M.; Hsu, C. S.; Kollin, E.; Millar, J.; Robbins, J.; Robbins, W.; Sherwood, R. D.; Tindall, P.; Fischer, D.; Meitzner, G. *Proc. Mater. Res. Soc. Boston* **1991**, *206*, 651.

28. Fleming, R. M.; Hessen, B.; Kortan, A. R.; Siegrist, T.; March, P.; Murphy, D. W.; Haddon, R. C.; Tycko, R.; Dabbagh, G.; Mujsce, A. M.; Kaplan, M.; Zahurak, S. M. *Proc. Mater. Res. Soc. Boston* **1990**, abstract.

29. Fleming, R. M.; Kortan, A. R.; Hessen, B.; Siegrist, T.; Thiel, F. A.; Marsh, P. M.; Haddon, R. C.; Tycko, R.; Dabbagh, G.; Kaplan, M. L.; Mujsce, A. M. *Phys. Rev. B* **1991**, *44*, 888.

30. Fleming, R. M.; Siegrist, T.; Marsh, P. M.; Hessen, B.; Kortan, A. R.; Murphy, D. W.; Haddon, R. C.; Tycko, R.; Dabbagh, G.; Mujsce, A. M.; Kaplan, M. L.; Zahurak, S. M. *Proc. Mater. Res. Soc. Boston* **1991**, *206*, 691.

31. Hawkins, J. M.; Lewis, T. A.; Loren, S. D.; Heath, J. R.; Saykally, R. J.; Hollander, F. J. *J. Chem. Soc. Chem. Commun.* **1991**, 775.

32. Heiney, P. A.; Fischer, J. E.; McGhie, R. A.; Romanow, W. J.; Denenstein, A. M.; McCauley, J. P., Jr.; Smith, A. B., III; Cox, D. E. *Phys. Rev. Lett.* **1991**, *66*, 2911.

Received September 19, 1991

Chapter 4

Solid C_{60}: Structure, Bonding, Defects, and Intercalation

John E. Fischer[1,2], Paul A. Heiney[1,3], David E. Luzzi[1,2], and David E. Cox[4]

[1]Laboratory for Research on the Structure of Matter,
[2]Materials Science Department, and [3]Physics Department,
University of Pennsylvania, Philadelphia, PA 19104–6272
[4]Physics Department, Brookhaven National Laboratory, Upton NY 11973

In this chapter, we review our X-ray and electron diffraction studies of pristine solid fullerite and the binary compounds obtained by doping to saturation with potassium, rubidium, or cesium. We describe an efficient and physically appealing method to model the high-temperature plastic crystal phase of solid C_{60}. A crystallographic analysis of the low-temperature orientationally ordered phase is presented. A native stacking defect is identified by electron diffraction, and its influence on powder X-ray profiles is explained. The compressibility of fullerite is consistent with van der Waals intermolecular bonding. Saturation doping with alkali metals leads to a composition M_6C_{60} (M is K, Rb, or Cs) and a transition of the C_{60} sublattice from face-centered cubic to body-centered cubic (fcc to bcc).

The recent discovery (*1*) of an efficient synthesis of C_{60} (buckminsterfullerene) has facilitated the study of a new class of molecular crystals (fullerites) based on these molecules (fullerenes). Solid C_{60} can be doped, or intercalated, with alkali metals, and the intercalated compounds show impressive values of electrical conductivity at 300 K (*2*) and superconductivity below 30 K (*3–5*). Atomic-resolution imaging, diffraction, and scattering techniques are being widely employed to study the structure, dynamics, bonding, and defects in this growing family of exciting new materials. The purpose of this chapter is to review the results obtained by the University of Pennsylvania–Brookhaven collaboration using X-ray and electron diffraction.

C_{60}: A Plastic Crystal

The first X-ray powder diffraction profile of solid C_{60} was analyzed in terms of a faulted hexagonal close-packed (hcp) lattice (*2*), consistent with ideal

0097–6156/92/0481–0055$06.00/0

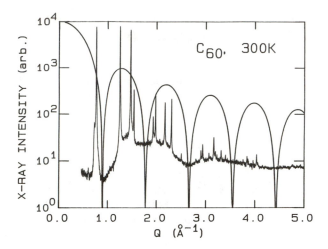

Figure 1. X-ray diffraction from pure C_{60} powder at 300 K. At this tempera-
ture there is no intermolecular orientational order, and the Bravais lattice is
face-centered cubic. The smooth curve is the square of the zero-order spherical
Bessel function, which represents the molecular form factor of a 3.53-Å radius
shell of charge. (Adapted from ref. 9.)

packing of spherical molecules and weak second-neighbor intermolecular interactions. A single-crystal study at 300 K showed (6) that the molecules are actually centered on sites of a rather perfect face-centered cubic Bravais lattice (fcc), but with a high degree of rotational disorder. The center-to-center distance between neighboring molecules was determined (3) to be 10.0 Å, consistent with a van der Waals (VDW) separation of 2.9 Å and a C_{60} diameter of 7.1 Å. Nuclear magnetic resonance spectroscopy clearly indicated the existence of dynamical disorder (presumably free rotation) that decreased with decreasing temperature (7, 8).

The data curve in Figure 1 is an X-ray powder profile of chromatographically pure, solvent-free C_{60}, taken with moderate resolution in a Debye–Scherrer (capillary) configuration (9). All the reflections can be indexed on an fcc cell with $a = 14.12$ Å, except for a weak broad feature superposed on the first strong reflection (described later). The first three reflections (111, 220, and 311) are observed to have comparable intensities, but there is no detectable 200 intensity (expected near 0.9 Å^{-1}). This finding is quite unusual for fcc Bravais lattices, but can be understood in terms of the peculiar X-ray form factor of orientationally averaged C_{60} molecules.

The NMR spectroscopic results show that the molecules are freely rotating at room temperature, so the ensemble-averaged charge density of each molecule is just a spherical shell. The Fourier transform of a uniform shell of radius R_o is $j_o(QR_o) = \sin(QR_o)/QR_o$, where j_o is the zero-order spherical Bessel function and $Q = 4\pi \sin \theta/\lambda$. The smooth, oscillatory curve in Figure 1 is a plot of this molecular form factor with R_o = 3.53 Å. This function has zeroes at values of Q corresponding to the expected h00 Bragg peaks with h even. All of the observed Bragg peaks occur within the "lobes" of the molecular form factor. As a consequence, the relative Bragg intensities are very sensitive to the combination of molecular radius and lattice constant.

A least-squares fit based on this model gives excellent agreement, indicating that the C$_{60}$ molecules exhibit little or no orientational order at 300 K (9). As probed by X-rays, the disorder could in principle be dynamic or static; the molecules could be spinning rapidly, as inferred from NMR spectroscopy (4, 5), or their icosahedral symmetry axes could exhibit no site-to-site orientational correlations. Either conjecture is consistent with the fact that a single-crystal refinement at 300 K fails to localize the polar and azimuthal angles of individual C atoms (6). The absence of detectable 200 intensity (and, indeed, of any h00 peaks with h even) is entirely due to the fact that $j_o(QR_o)$ has minima at the corresponding Q values when R_o = 3.5 Å and a = 14.12 Å.

An equally good fit to the 300-K profile can be obtained with a standard analysis based on an ad hoc space group that distributes the charge of 60 C atoms over 120 or 240 sites and incorporates sizable thermal disorder. The spherical shell approach has two advantages: fewer adjustable parameters and straightforward extension to orientationally ordered phases by including higher terms in a spherical harmonic expansion. The latter feature may be of great help in eventually reconciling temperature-dependent order parameters derived from NMR spectroscopic and diffraction data. Neutron scattering promises to provide additional information about the structure and dynamics of the plastic crystal phase of C$_{60}$ (10–12).

A Native Defect in Solid C$_{60}$

The origin of an anomalous powder diffraction feature has been established by transmission electron diffraction (13). This feature is consistently observed to some extent in powder profiles from solution-grown and sublimed samples, independent of temperature and hydrostatic pressure. Its rather large intensity in the very first samples of solid C$_{60}$ apparently prompted the original hcp indexing (1). Figure 2 shows part of a high-resolution powder profile. The sharp peak at Q = 0.771 Å^{-1} is the fcc 111 Bragg reflection. The enlarged curve clearly shows that the diffuse feature (if deconvoluted from the Bragg peak and the finite resolution) is sawtooth-shaped with a leading edge at about 0.72 Å^{-1} as indicated by the arrow. The leading edge could be indexed as the 100 reflection of a hexagonal close-packed lattice. With a(hcp) = 10.02 Å and c(hcp) = 16.36 Å, then Q(100) = 0.724 Å^{-1}, the fcc 111 becomes the hcp 002, and the trailing edge of the sawtooth is the stacking fault-broadened hcp 101 (0.818 Å^{-1} without faulting). Transmission electron microscopic (TEM)

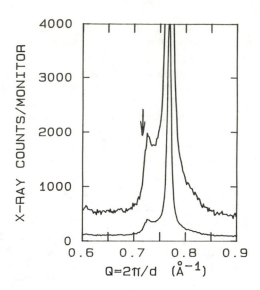

Figure 2. Expanded view of the fcc 111 reflection in pure C_{60}, showing super-position of the sharp Bragg peak on a sawtooth-shaped "diffuse" peak. The leading edge of the sawtooth corresponds to the fcc d-value 2/3, 2/3, 4/3 (see text).

analysis establishes that the sawtooth arises from powder-averaged rods of scattering that are most likely associated with growth defects.

Figure 3 shows a series of electron diffraction patterns recorded from C_{60} crystals sublimed directly onto amorphous carbon-coated grids. The sequence is for a [110] tilt axis, running from a [111] zone axis (Figure 3a), through intermediate tilt angles (Figures 3b–3d), ending with a [110] zone axis (Figure 3e). The 220 reflections are strong, but the 111 reflections are fairly weak, a result that is inconsistent with the X-ray (powder-averaged) results. This finding indicates that the X-ray results are streaked such that the [110] zone axis pattern intersects only a portion of the total intensity. The [111] zone axis pattern in Figure 3a exhibits the expected overall threefold symmetry. The strong reflections (large arrow) correspond to an interplanar spacing of 5.02 Å; therefore, they are 220 reflections. These lie closest to the transmitted beam spot, as expected for a normal fcc [111] pattern.

However, three other sets of reflections exist along $< 22\bar{4} >$ directions (small arrows). These occur at $Q = 0.72$ Å^{-1}, the same value as the leading edge of the sawtooth. With fcc indexing, this set corresponds to a Miller index of 2/3, 2/3, 4/3. These extra reflections persist at intermediate tilt angles (Figures 3b–3d), although their positions within the pattern vary systematically with tilt angle. As a guide to the eye, the expected positions of the

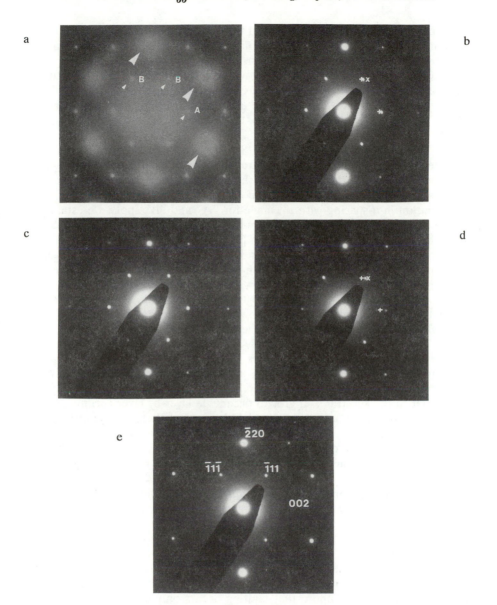

Figure 3. Electron diffraction patterns of C_{60} sublimed on holey carbon, recorded by tilting the crystal around the [110] tilt axis. Photos a–e correspond to tilt angles α away from [111] where α = 0, 10, 20, 30, and 35.3°, respectively. The extra reflections A and B undergo systematic displacements beginning in 2/3, 2/3, 4/3 positions in the [111] zone axis pattern (a) and moving horizontally in the photos to the [$\bar{1}$11]-type positions (A-type) or disappearing prior to reaching the 002 positions (B-type) in the [110] zone axis pattern (e). This behavior is consistent with [111] rods of diffuse scattering running through the $\bar{1}$11 reflection.

111 reflections in the [110] pattern and the 2/3, 2/3, 4/3 reflections in the [111] pattern are marked by × and + symbols, respectively. The reflections marked "A" move radially away from the transmitted beam spot with increasing tilt, whereas the reflections marked "B" move on a line between the "+" and "×" positions. In contrast to the continuous scattering vector variation of "2/3, 2/3, 4/3" reflections around the [111] zone axis, the 220 reflections that do not lie on the tilt axis behave normally, that is, they disappear abruptly as the crystal is tilted.

The behavior of the "A" and "B" spots indicates that these are not discrete reflections, but are due to streaks, or rods, of intensity through the reciprocal lattice. As the positions of these spots change with crystal tilt, their distance from the transmitted beam spot also changes. By analyzing in detail the variation in d-spacing with tilt angle for "A" and "B" spots, we established that the rods pass through (111) reciprocal lattice points parallel to [111] but not other equivalent directions, where [111] is the slow-growing crystal direction normal to the substrate. The closest approach of such a rod to the origin of reciprocal space is the point at which the rod is perpendicular to a line from the origin, namely (2/3, 2/3, 4/3), $Q = 0.726$ Å^{-1}, $d = 8.646$ Å. This result corresponds exactly to the leading edge of the sawtooth, which can now be understood as a variant of the well-known Warren line shape (*14*), modified on the high-Q side by the spherical form factor.

These unusual diffuse rods imply planar defects parallel to only one set of equivalent close-packed (111) planes. Random formation of classic hcp-type stacking faults in an fcc lattice would produce rods along all four equivalent < 111 > directions through each 111 reflection, not a single rod roughly perpendicular to the substrate. If growth occurs via layer-by-layer accumulation of single molecules on an initial close-packed layer, then there will be little driving force for a single fullerene to choose a B site versus a C site on an A-type layer. Layer-by-layer growth also implies that [111] will be a slow-growth direction, consistent with our observation of platelike grains with (111) faces parallel to the substrate.

We conclude that we are observing ..ABAB... faulting along a single [111] direction in an ...ABCABC.. sequence of close-packed planes. Krätschmer et al. (*1*) considered the converse case, namely ...ABC... faulting along the *unique* hexagonal [001] axis, for which reflections with indices $h - k = 3n \pm 1$ and $l \neq 0$ are broadened by an amount governed by the fault density. The other standard case, random ..ABAB.. faults along all four cubic < 111 > axes, results in the 111 remaining sharp, while reflections like 220, 311, etc., are broadened. Here we propose that faulting occurs only along a single < 111 > normal to the substrate, and it is straightforward to show that reflections whose indices satisfy $2h - k - l = 3n$ (n is an integer) should now be broadened. The stacking fault density in Figures 1 and 2 is apparently too small to reveal the consequences of this effect. We prepared a number of thin-film samples under a variety of sublimation conditions, some of which show large sawtooth amplitudes *and* large discrepancies in relative intensities that are not attributable to preferred orientation. A simplified analysis assuming a large density of faults along a single < 111 > direction gives qualitative agreement with the data.

There is clearly plenty of scope for additional work on defects in fullerenes. This work will be particularly important for the doped phases, in the context of optimizing the electronic and superconducting properties.

Orientationally Ordered C_{60}

Figure 4 compares high-resolution powder diffraction profiles of C_{60} measured at 300 and 11 K (same sample as Figures 1 and 2) (*15*). Compared to the 300-K fcc phase, the 11-K lattice constant has decreased by 0.13 Å, and many new peaks have appeared. These can all be indexed as simple cubic (sc) reflections with mixed odd and even indices, that is, forbidden fcc reflections. The crystal has therefore undergone a transition to a simple cubic structure, but the cube edge has not changed appreciably, so the basis must still consist of four molecules per unit cell, equivalent at high temperature but inequivalent at low temperature. Detailed measurements of the temperature-dependent integrated intensity of the sc 451 reflection reveal a weakly first-order transition at 249 K, which is also clearly observed in differential scanning calorimetry (*15*). These findings are all consistent with an orientational ordering transition, as confirmed by temperature-dependent NMR spectroscopy (*16*).

The 11-K profile is well-described by a standard crystallographic analysis based on orientationally ordered undistorted molecules (space group Pa3) (*17*) with some residual orientational disorder. Alternative assumptions that also remove the equivalence (displacements away from fcc Bravais lattice sites, quadrupolar molecular distortions into a "football" shape) gave poor results. The low-temperature sc lattice imposes severe constraints on possible models, because the equivalence of the x, y, and z axes and the corresponding molecular threefold axes must be maintained.

The model is constructed as follows. Four molecules, centered on fcc Bravais lattice sites, are oriented such that one of the 10 threefold icosahedral axes (normal to the pseudo-hexagonal faces) is aligned with one of the four < 111 > directions, and three mutually orthogonal twofold molecular axes are aligned with < 100 > directions. It follows that three other threefold axes are *also* aligned with the three remaining < 111 > axes. At this point there are no remaining rotational degrees of freedom; all molecules are equivalent and the structure is still fcc. The equivalence is now broken by rotating the four molecules through the same angle Γ but about *different* < 111 > axes. The specific rotation axes in Pa3 are [111], [11$\bar{1}$], [1$\bar{1}$1] and [$\bar{1}$11] for the molecules centered at (0, 0, 0), (1/2, 1/2, 0), (1/2, 0, 1/2), and (0, 1/2, 1/2), respectively. The best fit to the 11-K profile is shown by the solid curve in Figure 5 (dots are data points). The optimized C positions correspond to Γ = 26°. This is the same *crystal* structure as that of solid hydrogen, except that the degree of freedom represented by Γ is unnecessary for the cylindrical molecular symmetry of

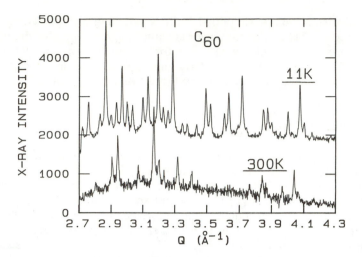

Figure 4. Portions of pure C_{60} X-ray powder profiles at 11 K (top curve) and 300 K (bottom curve), showing the onset of orientational order at low temperature. All reflections in the 300-K profile obey the fcc selection rule: h, k, and l either all odd or all even. New peaks with mixed odd and even indices appear in the 11-K profile, indicating that the four molecules per cube are no longer equivalent. (Adapted from ref. 15.)

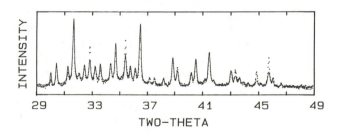

Figure 5. Least-squares fit to the 11-K data in Figure 4, based on a model of orientationally ordered molecules in space group Pa3. (Reproduced with permission from reference 17. Copyright 1991.)

H_2. Space group Pn3 is also compatible with icosahedral molecules oriented with respect to cubic axes; this condition gives a different set of rotation axes and a distinctly poorer fit. In particular, the presence of the 610 reflection is in disagreement with the systematic absences in space group Pn3. Low-temperature neutron diffraction data are also well-represented by the Pa3 structure (*12*).

An orientational ordering transition temperature, T_c = 249 K, is unusually high compared with, for example, the value of 20.4 K measured for CD_4

(*18*). However, C_{60} is a much larger molecule, with a rotational inertia several orders of magnitude larger than that of CD_4. Consequently, its motion will be much closer to the classical limit, and quantum tunneling will be substantially suppressed. Furthermore, the large number of equivalent orientations of a single fullerene molecule implies that the decrease in entropy per molecule from free rotation to fixed orientation is relatively small and results in an increased value of T_c. The temperature dependence of the rotational diffusion coefficient obtained from molecular dynamics simulations implies $T_c = 160$ K (*19*).

Compressibility and Intermolecular Bonding

The nature of intermolecular bonding is of considerable interest, both in its own right and as a clue to the electronic properties of fullerites and their derivatives. One expects a priori that the bonding will consist primarily of van der Waals interactions, analogous to interlayer bonding in graphite. Isothermal compressibility is a sensitive probe of interatomic–intermolecular bonding in all forms of condensed matter. We performed such an experiment on pure solid C_{60}, using standard diamond anvil techniques and powder X-ray diffraction (*9*). The three strongest peaks (111, 220, and 311) were recorded at 0 and 1.2 GPa and fitted to the predicted values for an fcc lattice with a as an adjustable parameter. We found that a decreased by 0.4 Å in this pressure range. Assuming no change in molecular radius, this decrease corresponds to a reduction in intermolecular spacing from 2.9 to 2.5 Å.

The a-axis compressibility, $-d(\ln a)/dP$, is 2.3×10^{-12} cm^2/dyne, essentially the same as the interlayer compressibility $-d(\ln c)/dP$ of graphite. Isothermal volume compressibilities, $-1/V(dV/dP)$ are 6.9, 2.7, and 0.18×10^{-12} cm^2/dyne for solid C_{60}, graphite, and diamond, respectively. (V is volume; P is pressure.) Clearly, fullerite is the softest all-carbon solid currently known. Another measurement at higher pressure shows that the compressibility decreases with increasing pressure, as expected (*20*). The dynamic range in both experiments was insufficient to reveal a possible onset of weak sc reflections with increasing pressure; one expects in principle that orientational freezing will occur at 300 K and elevated pressure. A full (temperature and pressure) study of C_{60} might help to identify the microscopic details of rotational dynamics.

At atmospheric pressure, the 2.9-Å van der Waals carbon diameter in solid C_{60} is considerably less than the 3.3-Å value characteristic of planar aromatic molecules and graphite. The observation of different diameters yet similar linear compressibilities between van der Waals-bonded planes can be explained qualitatively as follows. Intermolecular or interlayer separations in C_{60} and graphite, respectively, are determined by the balance between attractive and repulsive energies, and the corresponding compressibilities are defined by gradients of these energies. It is easy to show that the number of bonds per unit area parallel to a close-packed layer is at least 7 times smaller in C_{60} than in graphite. This disparity implies a large difference in total energies of the two solids, but does not *directly* account for the smaller spacing; if the close-packed layers were very stiff, the equilibrium spacing would be independent of bonds or area. However, the nature of the bonds is qualitatively different. The lobes of p_z charge in fullerene are normal to the spherical surface and probably remain nearly so in the solid. This condition permits a closer approach of neighboring $C_{60}s$ compared to graphite because the lobes extending into the intermolecular gap are "splayed out" with respect to a normal to the close-packed plane, rather than strictly normal to the plane as in graphite. Simple trigonometry shows that this effect alone can account for more than half the reduction in intermolecular spacing.

Alkali-Intercalated C_{60}

It has been shown (2) that M_xC_{60} (where M is an alkali metal) is metallic at 300 K over some range of x. Superconductivity occurs when $x = 3$ (5, 21), with onset temperatures of 18 and 29 K for M = K and Rb, respectively (3–5). These dramatic observations give further impetus to structural studies of doped fullerites. The rapidly growing literature on M_xC_{60} underscores the urgent need for detailed theoretical and experimental studies of the binary phase equilibria.

The design of our initial experiment (22) followed naturally from previous work on other guest–host systems [intercalated graphite (23), doped polymers (24)]. X-ray powder profiles were measured from equilibrium compositions of C_{60} doped to saturation with K, Rb, and Cs. Profiles that could be indexed as single phase were obtained by reacting pure, solvent-free, low-residue C_{60} powders (correlation length >1500 Å) with alkali vapor in evacuated glass tubes into which a large excess of alkali metal had been distilled, at temperatures on the order of 200–225 °C for at least 24 h. A gradient of 2–5 °C was maintained to avoid condensing metal onto the C_{60}. Shorter times or lower temperatures resulted in detectable amounts of unreacted C_{60}. Only a single "doped" phase was observed in all these experiments, either as a pure phase or in combination with the undoped fcc structure described earlier.

Figure 6 shows the 300-K powder diffractogram (dots) and a Rietfeld refinement (solid curve) for C_{60} doped to saturation with Cs. Similar profiles are obtained with K and Rb doping. All reflections can be indexed on a body-centered cubic lattice, with $a = 11.79$ Å; the corresponding values for K and Rb doping are 11.39 and 11.52 Å, respectively. Saturation doping therefore induces a significant rearrangement of C_{60} molecules with respect to the pure C_{60} fcc structure. The peak widths are not resolution-limited; the coherence length is 450 Å, about 1/3 the initial value. Diffuse scattering is not detectable from the sample.

An integrated intensity R factor of 4.3% is obtained for the refinement shown (space group Im3). All C–C distances were constrained to be equal,

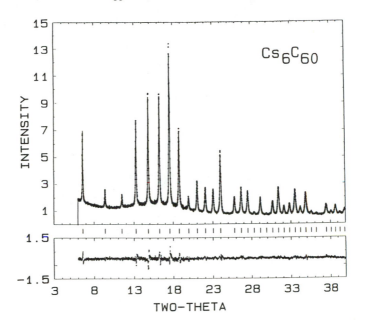

Figure 6. X-ray powder diffraction profile of Cs-doped C$_{60}$ (dots, $\lambda = 0.9617$ Å) and a Rietfeld refinement in space group Im3 (solid curve, lower panel shows data model). The bcc lattice constant is 11.79 Å, and the intensity R factor is 4.3%.

yielding a value 1.44 Å for this distance, slightly greater than the weighted mean of the NMR-determined values in pure C$_{60}$ (*25*). The R factor was not significantly improved by allowing two different bond lengths, or by an unconstrained refinement of all eight C positional parameters. Isotropic Debye–Waller factors yielded rms thermal amplitudes of 0.018 and 0.039 Å^2 for C and Cs, respectively. What is most significant is that the C$_{60}$ molecules in the saturation-doped phases exhibit complete orientational order; that is, all molecules have the same orientation with respect to the crystal axes. Doping-induced orientational "freezing" could be a consequence of the electrostatic field associated with electron transfer from M to C$_{60}$, or it could be a signature of orbital hybridization (partial covalent bonding).

Figure 7 shows a schematic view of a cube face, from which the essential features of the composite structure may be appreciated. Two equivalent C$_{60}$ molecules per cell are centered at (0, 0, 0) and (1/2, 1/2, 1/2) (the latter is omitted for clarity), oriented with twofold axes along cube edges. Twelve Cs atoms per cell are located at (0, 0.5, 0.28) and allowed permutations in space group Im3. These can be visualized as four-atom motifs centered on (1/2, 1/2, 0) and equivalent positions. The motif also has a twofold axis parallel to a cube edge. A typical motif lies in the {001} plane with atoms displaced $\pm0.28a$ along x and

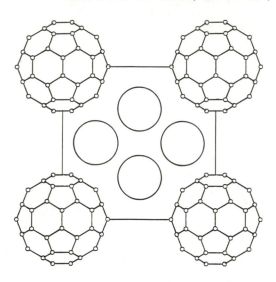

Figure 7. Schematic of the cube face normal to z, derived from Rietfeld refinement parameters. Large circles represent Cs^+ in scale with the cube edge. Small circles are C atoms, not to scale. An equivalent C_{60} molecule is centered at (1/2, 1/2, 1/2) (not shown). Cs coordinates (clockwise from top) are (0.5, 0.72, 0), (0.78, 0.5, 0), (0.5, 0.28, 0), and (0.22, 0.5, 0) with respect to an origin at the bottom left corner. Faces normal to x or y may be visualized by rotating the diagram $\pm 90°$. This is a consequence of molecular orientation with twofold axes along cube edges. (Reproduced with permission from reference 22. Copyright 1991 Macmillan Magazines Ltd.)

$\pm 0.22a$ along y from the face center. Each C_{60} is thus surrounded by 24 Cs atoms, and each Cs is in a distorted tetrahedral environment of four C_{60}s. The ideal stoichiometry is M_6C_{60}. The shortest distances between C_{60} centers are 9.86, 9.98, and 10.21 Å for K, Rb, and Cs, respectively. These bracket the 10.02-Å value in the undoped fcc phase (5–7). Near-neighbor C–Cs distances lie in the range 3.38–3.70 Å; the sum of van der Waals C and ionic Cs radii is 3.2 Å. All the Cs–Cs near-neighbor distances are 4.19 Å, considerably greater than the ionic diameter 3.34 Å. K_6C_{60} does not superconduct above 4.2 K; its 300-K electrical properties have not been measured.

Our initial experiments gave no evidence for other phases when the doping reaction was terminated before saturation, a finding that led us to believe that M_6C_{60} was the only stable phase. Subsequent work by other groups demonstrated the existence of a second doped phase M_3C_{60} (5) in which the fcc host sublattice is preserved and the M atoms occupy all available tetrahedral and octahedral vacancies (21). M_3C_{60} is apparently *the* superconducting phase. The fact that we did not observe it under nonequilibrium growth conditions is consistent with the small Meissner fraction observed in the initial discovery of superconductivity (*3*). We have recently confirmed the doped fcc structure, and the 18- and 29-K T_cs, for K_3C_{60} and Rb_3C_{60}, respectively.

Concluding Remarks

Further progress in understanding the structure, dynamics, and defect properties of C_{60} and doped phases requires that more attention be paid to detailed materials characterization. For example, a model proposed to explain the need for a three-step alkali metal doping procedure to optimize the superconducting fraction suggests that the crystallite size of the starting C_{60} might be an important parameter. Variable concentrations of stacking faults might affect the doping process via the diffusion rate and/or by affecting the shear displacements involved in the fcc–bcc transition. Most groups are working with chromatographically purified materials, so the issue of solvent removal and analysis merits further attention: In extreme cases, C_{60} can cocrystallize with organic solvents to produce entirely different phases (*26*). Finally, it seems quite unlikely that materials quoted as "pure C_{60}" contain nothing but C_{60}; HPLC analysis can rule out higher fullerenes, but other species may still be present, as suggested by substantial residues after thermogravimetric analysis above 900 °C in inert gas flows.

Acknowledgments

The results reported here could not have been obtained without the efforts of Arnold Denenstein, Andrew McGhie, and William Romanow, who made the soot and performed the toluene extractions; Nicole Coustel and John McCauley, who purified the C_{60}; Stefan Idziak, Stefan Kycia, Gavin Vaughan, Otto Zhou, and Qing Zhu, who materially contributed to the X-ray work; and X. Q. Wang and Debbie Ricketts-Foot, who carried out most of the TEM analyses. We are also grateful to Paul Chaikin (Princeton University) and Stan Tozer (Du Pont) who analyzed our samples for superconductivity, and to Ailan Cheng, Brooks Harris, Michael Klein, Gene Mele, Ward Plummer, and Amos Smith for stimulating conversations.

This research was supported variously by the National Science Foundation (NSF) Materials Research Laboratory Program DMR88–19885, by NSF DMR89–01219, and by the Department of Energy (DOE), DE–FC02–86ER45254 and DE–FG05–90ER75596. The National Synchrotron Light Source at Brookhaven is supported by DOE contract DEAC02–76CH00016.

References

1. Krätschmer, W.; Lamb, L. D.; Fostiropoulos, K.; Huffman, D. R. *Nature (London)* **1990,** *347,* 354.

2. Haddon, R. C.; Hebard, A. F.; Rosseinsky, M. J.; Murphy, D. W.; Duclos, S. J.; Lyons, K. B.; Miller, B.; Rosamilia, J. M.; Fleming, R. M.; Kortan, A. R.; Glarum, S. H.; Makhija, A. V.; Muller, A. J.; Eick, R. H.; Zahurak, S. M.; Tycko, R.; Dabbagh, G.; Thiel, F. A. *Nature (London)* **1991,** *350,* 320.

3. Hebard, A. F.; Rosseinsky, M. J.; Haddon, R. C.; Murphy, D. W.; Glarum, S. H.; Palstra, T. T. M.; Ramirez, A. P.; Kortan, A. R. *Nature (London)* **1991,** *350,* 600.

4. Rosseinsky, M. J.; Ramirez, A. P.; Glarum, S. H.; Murphy, D. W.; Haddon, R. C.; Hebard, A. F.; Palstra, T. T. M.; Kortan, A. R.; Zahurak, S. M.; Makhija, A. V. *Phys. Rev. Lett.* **1991,** *66,* 2830.

5. Holczer, K.; Klein, O.; Huang, S. M.; Kaner, R. B.; Fu, K. J.; Whetten, R. L.; Diederich, F. *Science (Washington, D.C.)* **1991,** *252,* 1154.

6. Fleming, R. M.; Siegrist, T.; March, P. M.; Hessen, B.; Kortan, A. R.; Murphy, D. W.; Haddon, R. C.; Tycko, R.; Dabbagh, G.; Mujsce, A. M.; Kaplan, M. L.; Zahurak, S. M. In *Clusters and Cluster-Assembled Materials*; Averback, R. S.; Bernholc, J.; Nelson, D. L., Eds.; Materials Research Society Symposium Proceedings 206, Materials Research Society: Pittsburgh, PA, 1991; p 691.

7. Tycko, R.; Haddon, R. C.; Dabbagh, G.; Glarum, S. H.; Douglass, D. C.; Mujsce, A. M. *J. Phys. Chem.* **1991,** *95,* 518.

8. Yannoni, C. S.; Johnson, R. D.; Meijer, G.; Bethune, D. S.; Salem, J. R. *J. Phys. Chem.* **1991,** *95,* 9.

9. Fischer, J. E.; Heiney, P. A.; McGhie, A. R.; Romanow, W. J.; Denenstein, A. M.; McCauley, J. P., Jr.; Smith, A. B., III *Science (Washington, D.C.)* **1991,** *252,* 1288.

10. Cappelletti, R. L.; Copley, J. R. D.; Kamitakahara, W. A.; Li, F.; Lannin, J. S.; Ramage, D. *Phys. Rev. Lett.* **1991,** *66,* 3261.

11. Li, F.; Ramage, D.; Lannin, J. S.; Conceicao, J. *Phys. Rev. Lett.* submitted, 1991.

12. Copley, J. R. D.; Neumann, D. A.; Cappelletti, R. L.; Kamitakahara, W. A.; Prince, E.; Coustel, N.; McCauley, J. P., Jr.; Maliszewskyj, N. C.; Fischer, J. E.; Smith, A. B., III; Creegan, K. M.; Cox, D. M. Proceedings of an International Conference on Neutron Scattering, Oxford, United Kingdom, *Physica B,* in press, 1991.

13. Luzzi, D. E.; Fischer, J. E.; Wang, X. Q.; Ricketts-Foot, D. A.; McGhie, A. R.; Romanow, W. J. *J. Mater. Res.,* in press, 1991.

14. Warren, B. E. *Phys. Rev.* **1941**, *59*, 693.

15. Heiney, P. A.; Fischer, J. E.; McGhie, A. R.; Romanow, W. J.; Denenstein, A. M.; McCauley, J. P., Jr.; Smith, A. B., III *Phys. Rev. Lett.* **1991**, *66*, 2911.

16. Tycko, R.; Dabbagh, G.; Fleming, R. M.; Haddon, R. C.; Makhija, A. V.; Zahurak, S. M. *Phys. Rev. Lett.* **1991**, *67*, 1886.

17. Sachidanandam, R.; Harris, A. B. *Phys. Rev. Lett.* **1991**, *67*, 1467.

18. Press, W.; Kollmar, A. *Solid State Commun.* **1975**, *17*, 405; Silvera, I. F. *Rev. Mod. Phys.* **1980**, *52*, 393.

19. Cheng, A.; Klein, M. L. *J. Phys. Chem.*, in press, 1991.

20. Duclos, S.; Brister, K.; Haddon, R. C.; Kortan, A. R.; Thiel, F. A. *Nature (London)* **1991**, *351*, 380.

21. Stephens, P. W.; Mihaly, L.; Lee, P. L.; Whetten, R. L.; Huang, S-M.; Kaner, R. B.; Diederich; F.; Holczer, K. *Nature (London)* **1991**, *351*, 632.

22. Zhou, O.; Fischer, J. E.; Coustel, N.; Kycia, S.; Zhu, Q.; McGhie, A. R.; Romanow, W. J.; McCauley, J. P., Jr.; Smith, A. B., III *Nature (London)* **1991**, *351*, 462.

23. Fischer, J. E. In *Chemical Physics of Intercalation*; Legrand, A. P.; Flandrois, S., Eds.; NATO ASI Series B172; Plenum: New York, 1987; p 59.

24. Heiney, P. A.; Fischer, J. E.; Djurado, D.; Ma, J.; Chen, D.; Winokur, M. J.; Coustel, N.; Bernier, P.; Karasz, F. E. *Phys. Rev. B* **1991**, *44*, 2507.

25. Yannoni, C. S.; Bernier, P.; Bethune, D. S.; Meijer, G.; Salem, J. R. *J. Am. Chem. Soc.* **1991**, *113*, 3190.

26. Fleming, R. M.; Kortan, A. R.; Hessen, B.; Siegrist, T.; Thiel, F. A.; Marsh, P. M.; Haddon, R. C.; Tycko, R.; Dabbagh, G.; Kaplan, M. L.; Mujsce, A. M. *Phys. Rev. B* **1991**, *44*, 888.

Received August 16, 1991

Chapter 5

Conductivity and Superconductivity in Alkali Metal Doped C_{60}

R. C. Haddon, A. F. Hebard, M. J. Rosseinsky, D. W. Murphy,
S. H. Glarum, T. T. M. Palstra, A. P. Ramirez, S. J. Duclos, R. M. Fleming,
T. Siegrist, and R. Tycko

AT&T Bell Laboratories, Murray Hill, NJ 07974–2070

Solid C_{60} undergoes doping with alkali metal vapors to produce intercalation compounds that are conductors. During the doping process the predominant phases present are C_{60}, A_3C_{60}, and A_6C_{60}. The A_3C_{60} compounds are formed from C_{60} by occupancy of the interstitial sites of the face-centered cubic (fcc) lattice. These phases constitute the first three-dimensional organic conductors and for A = K or Rb, the A_3C_{60} compounds are superconductors. In this chapter we summarize the current status of the research on these new materials, including some of the physical properties that have been measured. The structure, conductivity, magnetism, microwave loss, and Raman and electronic structures are discussed.

Organic conductors depend on the presence of π-electrons for their electronic transport properties. In extended systems such as polymers and graphite (*1–3*), the π-system directly provides a conducting pathway, whereas in molecular systems (*3–5*) the transport properties depend on the overlap between the π-orbitals on adjacent molecules. The nature of the overlap is a crucial feature of the properties of molecular conductors, and the directionality of the π-orbitals exerts a profound effect on the resultant electronic properties. For planar hydrocarbon arene (Ar)-based organic conductors (Ar is fluoranthene, perylene, or naphthalene), the ramifications are particularly obvious because the π-orbitals possess a well-defined directionality, that is, perpendicular to the molecular plane. In order to maintain favorable intermolecular overlap, these systems therefore stack in a perpendicular manner. This behavior is well known for the Ar_2PF_6 compounds, which show quite high conductivity together with highly anisotropic one-dimensional electronic properties (*6*). The anisotropy follows for the same reason: As the conduction band is composed almost

0097–6156/92/0481–0071$06.00/0

exclusively of π-orbitals that are directed along the stack, there is virtually no component of overlap between stacks.

The situation changes with the introduction of heteroatoms, such as the chalcogenides, which possess two lone-pair orbitals that may be considered as approximate sp^3 hybrids or as an sp^2 plus p orbital. The structures of some of the sulfur- and selenium-based charge-transfer salts reveal that the heteroatoms allow interactions both along a stack and between neighboring stacks in the crystal lattice. This conclusion is particularly obvious in the $(TMTSF)_2X$ (X is ClO_4, PF_6, etc., and TMTSF is tetramethyltetraselenafulvalene), salts that provided the first organic superconductors (7). In these compounds, there is still considerable anisotropy, but the interactions between the stacks are sufficient to inhibit the formation of an insulating ground state in the perchlorate. Nevertheless, this class of compounds is sufficiently close to the one-dimensional crossover, so that some members show spin density wave ground states at atmospheric pressure (4).

In many of the $(ET)_2X$ (ET is bis(ethylenedithio)tetrathiofulvalene) salts the stacking is no longer favored, and pairs of molecules are arranged end-on in a planar array (3–5). Within this two-dimensional sheet the interactions between the molecular pairs are almost isotropic as reflected in the structures and in band structure calculations (4, 5). A number of the $(ET)_2X$ salts are superconductors (8) and do not show evidence for the insulating ground state that is characteristic of low-dimensional compounds. The compounds κ-$(ET)_2Cu(NCS)_2$ (9) and κ-$(ET)_2Cu[N(CN)_2]Br$ (10) show superconducting transition temperatures of 10.4 and 12.4 K, respectively. These materials may be prepared as high-quality crystals that show excellent superconducting properties as inferred from microwave loss experiments (11). Although isotropic in two dimensions, this class of compounds is highly anisotropic in the third direction, and in this respect they resemble the high-T_c (critical temperature) cuprate superconductors.

As a result of the directionality of the π-orbitals, there have been no examples of three-dimensional isotropic molecular conductors. This constraint is in fact not imposed by the π-orbitals themselves, but by the planarity of the molecular framework that is usually thought necessary for delocalized π-bonding. However, a number of molecules have been reported (12) to show marked nonplanarity and are clearly aromatic.

With the observation (13) and synthesis (14) of the fullerenes, this point became irresistible. The availability of these systems provided a set of molecules with π-orbitals radiating in all directions. If ever a three-dimensional (3-D) electronic molecular solid could be realized, then these molecules represented the ideal vehicle.

The large size and high electron affinity (15–18) of the fullerenes provided a means to test this approach. By choosing a small dopant ion we reasoned that it should be possible to intercalate the fullerene crystal without disrupting the network of contacts between the spheroids and thereby generate the first 3-D isotropic organic conductor (19). The face-centered cubic (fcc) lattice (20) of C_{60} (buckminsterfullerene) provides three interstitial sites per

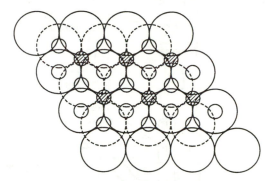

Figure 1. Diagram of two close-packed planes of spheres (large circles) show-ing the two types of interstitial sites. The large circles represent the C_{60} molecules, and the smaller circles represent the tetrahedral (open), and octahedral (cross-hatched) lattice sites that are available to the intercalants.

molecule, two tetrahedral and one octahedral, that are of sufficient size to be occupied by alkali metal cations (Figure 1) (*19*). These considerations led us to attempt the alkali metal doping of thin films of C_{60} and C_{70}. As we wished to explore highly reduced states of these molecules, we constructed a high-vacuum apparatus that would allow us to measure conductivities in situ (Figure 2a) (*19*).

Thin Films

Thin films of C_{60} and C_{70} were deposited on a variety of substrates and charac-terized with a number of techniques (*19, 21*). The C_{70} samples used as source material in film growth were obtained by reversed-phase column chromatogra-phy of crude fullerite (*14*) on octadecylsilanized silica using 40:60 toluene–2-propanol as the eluant, whereas the C_{60} samples were obtained by chromatog-raphy of fullerite on alumina using 5:95 toluene–hexane as eluant (*22*). The purities of the C_{60} and C_{70} were checked by proton NMR spectroscopy and high-pressure liquid chromatography (HPLC) using UV detection. High-quality films of C_{60} (C_{70}) of thickness 200–1000 Å were deposited by high-vacuum sublimation from an alumina crucible regulated at 300 (350) °C at a pressure of 1.5×10^{-6} torr (200×10^{-6} Pa). The C_{60} (C_{70}) films appeared smooth and pale yellow (magenta) to the eye. X-ray powder diffraction experi-ments on the C_{60} films indicate the presence of crystalline domains with coher-ence lengths on the order of 60 Å (*21*). Infrared spectra of the C_{60} films depo-sited on KBr substrates showed the four characteristic absorptions of C_{60} (*14*) with no evidence of contaminants.

For the conductivity measurements, the films were deposited on glass slides that had been precoated with 1000-Å stripes or pads of evaporated sil-

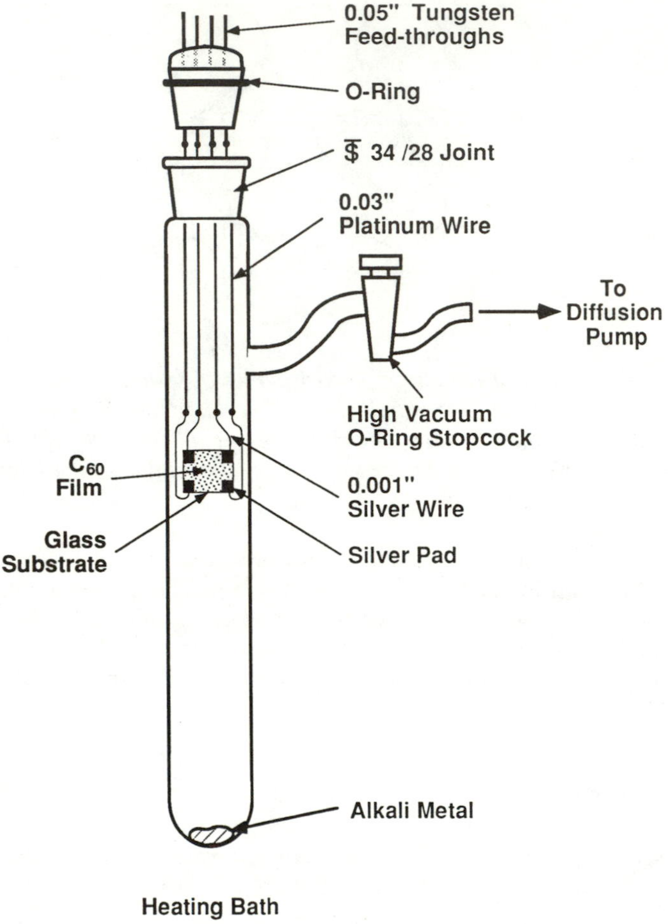

0.05" Tungsten
Feed-throughs
O-Ring
$ 34 /28 Joint
0.03"
Platinum Wire
To
Diffusion
Pump
High Vacuum
O-Ring Stopcock
C60
Film
0.001"
Silver Wire
Glass
Substrate
Silver Pad
Alkali Metal
Heating Bath

Figure 2a. Apparatus for measuring conductivities of C_{60} and C_{70} films as a function of vapor phase doping with alkali metals at room temperature. (Reproduced with permission from reference 19. Copyright 1991 MacMillan Magazines Ltd.)

ver metal (Figure 2). The measured two-probe resistance in the pristine films was greater than 10^{10} ohms, a result that implied a conductivity of less than 10^{-5} S/cm. Subsequent work has shown that pristine C_{60} is a semiconductor with a bandgap of 1.7 eV (*23, 24*) and a dielectric constant of 4.4 ± 0.2 (*21*).

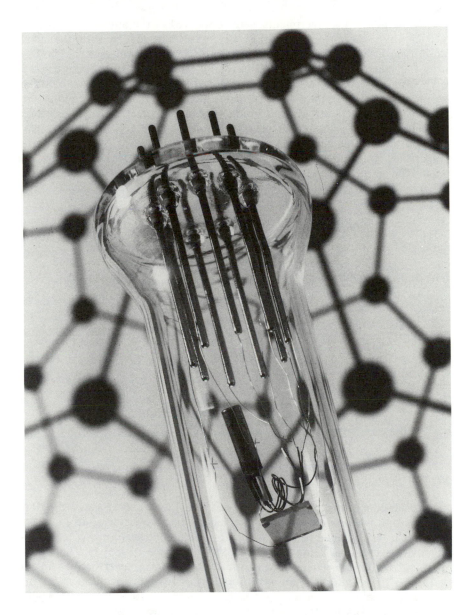

Figure 2b. Apparatus for measuring conductivities of C$_{60}$ *and C*$_{70}$ *films as a function of vapor phase doping with alkali metals at low temperature. (Reproduced with permission from reference 19. Copyright 1991 MacMillan Magazines Ltd.)*

Conducting and Superconducting Thin Films

The initial conductivity experiments were conducted in the apparatus shown in Figure 2a (*19*). After film mounting, the vessel was loaded with the alkali metal dopant in a dry box, before evacuation with a diffusion pump. The bottom of the apparatus was immersed in an oil bath, and the temperature was slowly raised until conductivity in the film could be detected. All of the doping experiments showed qualitatively the same behavior: First the conductivity increased by several orders of magnitude, and then it decreased, usually to a point below our threshold of detection. Doping of the C_{60} films with alkali metals led to magenta films with fairly good surface quality, although cesium doping visibly roughened the film. The C_{70} film showed little color change on doping. The results for the combinations that were originally tested are summarized in Table I (*19*).

Table I. Alkali Metal Doped Films of C_{60} and C_{70}

Film	Dopant	Bath Temperature (°C)	Maximum Conductivity (S/cm)
C_{60}	Li	a	10
C_{60}	Na	180	20
C_{60}	K	130	500
C_{60}	Rb	120	100
C_{60}	Cs	40	4
C_{70}	K	120	2

[a]Lithium metal in contact with Kovar container and flame-heated.
SOURCE: Reproduced with permission from reference 19. Copyright 1991 MacMillan Magazines Ltd.)

In subsequent ultrahigh vacuum studies (*25*), we reinvestigated the doping of C_{60} films with potassium from a molecular-beam effusion source. In this experiment it was possible to determine the stoichiometries (x in K_xC_{60}) at the conductivity extrema by Rutherford back-scattering spectroscopy. The results support our earlier inferences drawn from in situ Raman spectroscopy on the doped films (*19*). Conductivities were found to increase exponentially with initial potassium exposure, reaching a maximum value of 450 S/cm at $x = 3.00 \pm 0.05$ (Figure 3). Beyond this point the conductivity decreased with further K exposure until the stoichiometry $x = 6.00 \pm 0.05$ was reached. Of the C_{60} dopants, potassium gave rise to the highest conductivities in our initial experiments, and it is now known that this combination is relatively easy to optimize and also gives rise to superconductivity at the composition K_3C_{60} (*26*).

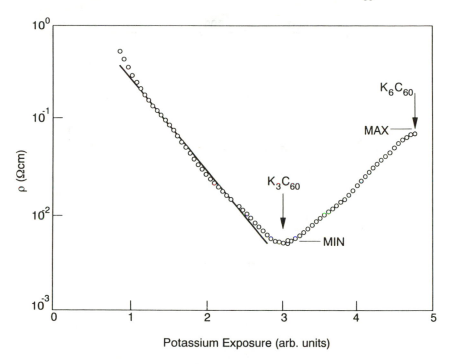

Figure 3. Time dependence of the resistivity of a K$_x$C$_{60}$ film during exposure to a potassium molecular beam in UHV at an ambient temperature of about 347 K (25).

In our initial work (*19*) we followed the doping of the films by in situ Raman spectroscopy. Figure 4a shows the Raman spectrum of a superconducting rubidium-doped film, whereas Figure 4b shows the insulating fully doped film (Duclos, S. J., unpublished). At the superconducting composition the high frequency A_g mode of C$_{60}$ is shifted to about 1445 cm^{-1}, and in the highly doped insulating state it moves to about 1430 cm^{-1}. On the basis of our previous correlation (*19*), we assign these species to Rb$_3$C$_{60}$ and Rb$_6$C$_{60}$. Addition of electrons to the C$_{60}$ framework is expected to soften the bond-stretching modes as the added electrons enter antibonding molecular orbitals (*15*). The doped C$_{60}$ film rapidly returned to its pale yellow color on exposure to the atmosphere, and a Raman spectrum showed that the line due to neutral C$_{60}$ had been restored. These observations suggest that the molecular integrity of C$_{60}$ is maintained on doping and that the process is chemically reversible (*19*).

In our initial report on superconductivity in the alkali metal fullerides (*26*), we were able to observe zero resistivity in a potassium-doped C$_{60}$ film. The experiment was performed in the all-glass apparatus shown in Figure 2b, which was sealed under a partial pressure of helium before doping and immersion in the Dewar flask. The room temperature resistivity of the K$_x$C$_{60}$ thin film was 5×10^{-3} ohm cm and increased by a factor of 2 on cooling the film to

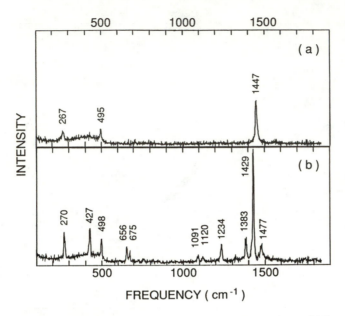

Figure 4. In situ Raman spectra of a C_{60} film taken during rubidium doping: (a) superconducting and (b) insulating.

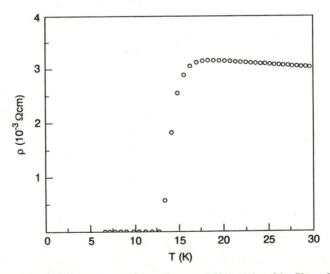

Figure 5. Temperature dependence of the resistivity of a thin film of K_3C_{60}.

20 K. Below 16 K, the resistivity began to fall, and zero resistivity ($<10^{-4}$ of the normal state) was observed at 5 K (26). The resistivity of the K_3C_{60} thin film shown in Figure 5 reaches zero at about 13 K (Palstra, T. T. M., unpublished).

We have also observed superconductivity in doped C_{60} films by microwave loss experiments on films grown along the walls of electron spin

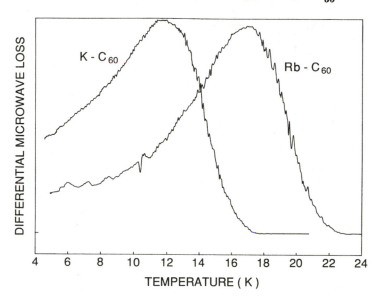

Figure 6. Microwave loss as a function of temperature for potassium- and rubidium-doped C_{60} films in a static field of 20 Oe.

resonance (ESR) tubes (Glarum, S. H., unpublished). By sealing a physically separated capillary of the appropriate alkali metal in the ESR tube with the preformed film, it was possible to carry out doping experiments leading to the full range of stoichiometries. Representative traces of the microwave loss versus temperature for potassium- and rubidium-doped C_{60} films are shown in Figure 6. For the rubidium-doped film in particular, the T_c is noticeably lower than that observed in bulk samples (discussed later). We were able to localize the superconductivity in these thin-film samples with an accuracy of about 1 mm, and this capability allowed us to perform in situ Raman studies on the doped films. On the basis of the Raman shift, we inferred superconducting stoichiometries of K_3C_{60} and Rb_3C_{60} in the films shown in Figure 6. The only compositions observed in these experiments were C_{60}, A_3C_{60}, and A_6C_{60}.

Superconducting Bulk Materials

As noted in our initial report (26), only the K–C_{60} combination showed super-conductivity under the relatively mild heating treatments used in the microwave loss screening experiments on bulk powders. The d.c. magnetization of a bulk sample of nominal composition K_3C_{60} that was prepared at a temperature of 200 °C showed a well-defined Meissner effect below 18 K (Figure 7) (26). The initial experiments showed a flux exclusion in the zero-field cooled curve corresponding to a 1% volume fraction. More recently we have observed flux exclusions of 40% in powder samples of K_3C_{60} (Ramirez, A. P., unpublished).

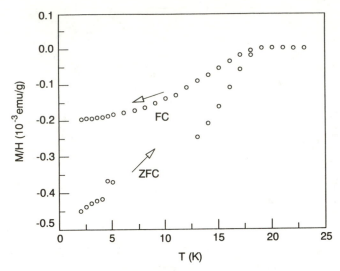

Figure 7. Temperature dependence of the magnetization of a K_xC_{60} sample. (Reproduced with permission from reference 26. Copyright 1991 MacMillan Magazines Ltd.)

a.c. magnetization studies of potassium-doped C_{60} samples prepared at the University of California at Los Angeles (UCLA) have been reported (*27, 28*), showing 40% (powder) and 100% (pressed powder) flux expulsion at the composition K_3C_{60} and with $T_c = 19.3$ K.

As discussed, microwave loss studies on rubidium-doped C_{60} films showed superconducting onsets of 23–26 K, and a bulk sample of nominal composition Rb_3C_{60} that was prepared at 400 °C showed a Meissner fraction of 1% with a superconducting transition temperature of 28 K (*29*). Concurrent experiments by the UCLA group (*27*) found a flux exclusion of 7% and a T_c of 30 K at the same composition by a.c. magnetization experiments. The d.c. magnetization of an Rb_3C_{60} sample with a Meissner fraction of 35% is shown in Figure 8 (Ramirez, A. P., unpublished).

It is interesting to compare the microwave loss data on thin films (Figure 6) with our original reports on bulk materials characterized with the same technique (*26, 29*). The microwave signal rises much more steeply in the bulk samples, and the measured transition temperatures are higher, particularly in the case of rubidium doping.

The ^{13}C NMR spectrum of a sample of nominal composition $K_{1.5}C_{60}$ is shown in Figure 9 (*30*). The sharp upfield resonance is due to neutral C_{60}, whereas the broad resonance has been assigned to K_3C_{60}. Clearly, this composition undergoes phase separation, and for K_xC_{60} compositions with $0 < x < 3$, the only two stable phases are C_{60} and K_3C_{60} (*30*).

Structural Studies of Alkali Fullerides

In the solid state, pristine C_{60} is arranged in a face-centered cubic array of rigid spheres with lattice constant $a = 14.2$ Å (*20*) with two tetrahedral and one

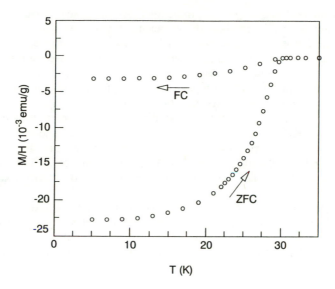

Figure 8. *Temperature dependence of the magnetization of a Rb$_x$C$_{60}$ sample.*

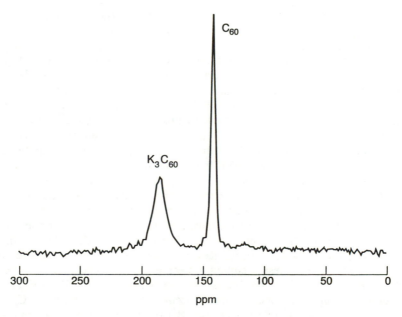

Figure 9. *^{13}C NMR spectrum of a sample of nominal composition of K$_{1.5}$C$_{60}$* (30).

octahedral interstitial site per C_{60} molecule. As discussed in our initial paper (19), we intended to intercalate these vacancies with the dopant counterions. Providing there was no drastic change in crystal structure, therefore, this model could account for the incorporation of a maximum of three dopant ions per C_{60} molecule and would lead to compositions of A_xC_{60} with $x = 0$–3. Our Raman studies supported the composition A_3C_{60} for the highly conducting and superconducting phase in this system (19, 26, 29).

Recent work (28) has demonstrated that the superconductivity is associated with the K_3C_{60} composition that has a face-centered cubic structure derived from that of C_{60} by incorporating potassium ions into all of the octahedral and tetrahedral interstices of the host lattice. There is a small expansion of the lattice constant from 14.11 to 14.28 Å on incorporation of the potassium intercalant.

We deduced from Raman spectroscopy that the insulating doped composition corresponded to A_6C_{60}, and this stoichiometry cannot be accommodated in the essentially unperturbed C_{60} lattice. Recent work (31) has now shown that the C_{60} structure is transformed to body-centered cubic at the composition A_6C_{60} for A = K and Cs.

Band Structure Calculations

We originally attributed the conductivity induced in these materials to energy bands composed of C_{60} π-orbitals that become populated in the doping process (19). As these bands fill at high doping levels the material becomes insulating. Based on the energy level structure of C_{60} we suggested that the conductor–insulator transition occurs at the point where each molecule has accepted about six electrons, and has thereby filled the t_{1u} level (15).

We have examined the electronic structure of solid C_{60} with extended Hückel theory (EHT) band structure calculations (32). EHT band structures have been extensively used in qualitative studies of the electronic structure of organic and inorganic conducting solids (32). Although the calculations were carried out for the neutral material, the similarities found between the the crystal structure of pristine C_{60} and the K_3C_{60} phase (28) suggest that the results should have some relevance for the conducting and superconducting materials.

The C_{60} molecular structure was taken from an MNDO (modified neglect of differential overlap) optimization that gave bond lengths of 1.400 and 1.474 Å for the free molecule (33). We first tested the EHT method for free C_{60} and found that the σ-orbitals were calculated to lie too high relative to the π-orbitals. The highest occupied molecular orbital of C_{60} is of π-character and lies about 3 eV above the first set of σ-orbitals (34, 35). With EHT theory the highest occupied σ- and π-orbitals occur very close together in C_{60}. This artifact of the calculations occurred with the original EHT parameters (36) and with those that we have employed (37).

This problem is not restricted to C_{60}. In the original EHT paper on hydrocarbons (*36*) it was noted that, in calculations on aromatic compounds, the σ and π levels were interspersed. The ionization energies (in electron volts) of the highest occupied orbitals of benzene were found by EHT and from photoelectron spectroscopy (PS) (*38*) to be

Orbital	EHT	PS
$e_{1g}(\pi)$	12.8	9.3
$e_{2g}(\sigma)$	12.8	11.4
$b_{2u}(\sigma)$	14.3	—
$a_{2u}(\pi)$	14.5	12.1

Because of this artifact of the EHT calculations, we exclude the valence levels from our discussion of solid C_{60}. We used modified EHT parameters (*37*) in our calculation and a double-zeta basis set (*39*). The π-orbitals themselves appear to be reasonably well described, and the calculated energy gap between the π-orbitals in the C_{60} molecule is 2.1 eV, which may be compared with the experimental value of 1.7 eV (*23, 24*).

We used the face-centered cubic lattice for solid C_{60}, with lattice constant $a = 14.197$ Å (*20*), and oriented the C_{60} molecules in the unit cell so that the final space-group was $Fm\bar{3}$. A close-packed plane of the idealized structure used in the band structure calculations is shown in Figure 10. The density of states (DOS) calculation was carried out by choosing a weighted grid of k points in the asymmetric unit of the conventional unit cell (*40*).

The energies of the two lowest vacant bands arising from the fcc lattice of C_{60} molecules are shown in Figure 11. The conduction band in solid C_{60} is derived from the molecular t_{1u} level, whereas the next lowest band arises from the t_{1g} level (*15*). The calculated dispersion of the conduction band is similar to that obtained recently in local-density calculations (*41*).

The calculated DOS is shown in Figure 11, and although the π-derived bands are in qualitative agreement with experiment and other theoretical treatments (*34, 35, 41*), it is apparent that the σ-orbitals are also calculated to lie too high relative to the π-orbitals in solid C_{60}. The energy gap between π-orbitals is calculated as 2 eV in solid C_{60}, which may be compared with the experimental value of 1.7 eV.

At the simplest level the conductivity in A_xC_{60} compounds would be interpreted as arising from the population of the conduction band by electrons donated to the C_{60} molecules by the intercalants. Although this picture is broadly supported by recent photoemission experiments on the potassium doping of C_{60} films (*42, 43*), the whole band becomes visible at low doping levels, and this behavior may be due to nonrigid band effects (*42*). However, the transport properties of the films are sensitive to film depth up to 1000 Å (*25*), and measurements on the film surface may not reflect the bulk properties. The films also suffer from a lack of long-range order (*21, 44*).

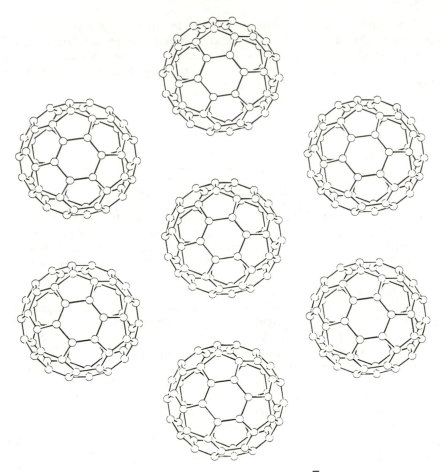

Figure 10. Close-packed plane from the idealized Fm$\bar{3}$ structure adopted for crystalline C$_{60}$.

At the composition A$_3$C$_{60}$ the conduction band would be half-filled, and this corresponds to the stoichiometry of maximum conductivity. The calculated width of this band is about 0.5 eV, and the DOS at the Fermi level is 13 states/eV C$_{60}$. Beyond this doping level the structure changes, but the insulating behavior at A$_6$C$_{60}$ in the bcc structure presumably arises from the filling of a similar band formed from the molecular t_{1u} level that can hold a maximum of six electrons (*15*).

The possibility of accessing other band-fillings within the fcc-derived C$_{60}$ structure is worth considering. Trivalent intercalants (T) would be of particular interest; at the composition T$_1$C$_{60}$, the band-filling would correspond to that in the A$_3$C$_{60}$ compounds, presumably with the intercalant occupying the octahedral site in the lattice. At the composition T$_3$C$_{60}$ there would be the

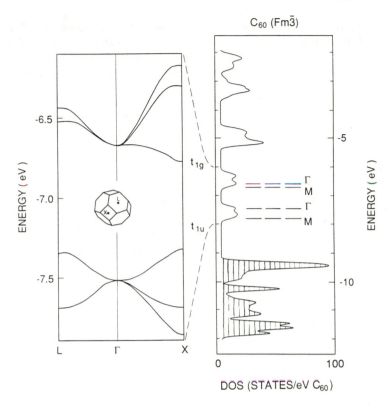

Figure 11. Calculated dispersions for the vacant t$_{1u}$- *and* t$_{1g}$-*derived bands, and density of states for fcc C$_{60}$. The labels M and Γ refer to the energy levels in the free molecule and in the solid at the zone center.*

possibility of injecting carriers into the t_{1g}-derived band, which we have previously estimated to be accessible under certain circumstances (*15*). This band is a little wider than the t_{1u}-derived band, 0.6 versus 0.5 eV. The DOS at the Fermi levels for half-filling are calculated to be 13 and 8.5 states/eV C$_{60}$ for the t_{1u}- and t_{1g}-derived bands, respectively.

Current Status

The normal-state conductivity of the alkali fullerides seems to be reasonably well accounted for within a simple band picture, and our original goal of producing a 3-D organic conductor (*19*) has apparently been realized. The metallic state occurs at the half-filled band composition, which has usually been associated with insulating electronic instabilities in the field of organic conductors as a result of their low dimensionality.

The nature of the superconductivity remains an open question, although a number of theoretical treatments have appeared (*45–48*). The 3-D nature of the A_3C_{60} compounds obviates many of the insulating ground states that usually compete with superconductivity at low temperatures. The dependence of T_c on intercalant seems to be accounted for by variations in the density of states at the Fermi level (*29, 49*).

Clearly, a number of possibilities exist for the synthesis of new intercalates of the fullerenes.

Acknowledgements

We thank D. N. E. Buchanan, E. E. Chaban, P. H. Citrin, G. Dabbagh, R. H. Eick, A. T. Fiory, G. P. Kochanski, A. R. Kortan, K. B. Lyons, A. V. Makhija, B. Miller, A. J. Muller, K. Raghavachari, J. E. Rowe, J. M. Rosamilia, F. A. Thiel, G. K. Wertheim, W. L. Wilson, and S. M. Zahurak for valuable contributions to this work.

References

1. *Handbook of Conductive Polymers*; Skotheim, T. A., Ed.; Marcel Dekker: New York, 1986.

2. *Chemical Physics of Intercalation*; Legrand, A. P.; Flandrois, S., Eds.; Plenum: New York, 1987.

3. Ferraro, J. R.; Williams, J. M. *Introduction to Synthetic Electrical Conductors*; Academic: London, 1987.

4. Ishiguro, T.; Yamaji, K. *Organic Superconductors*; Springer-Verlag: Berlin, 1990.

5. Kresin, V. Z.; Little, W. A. *Organic Superconductivity*; Plenum: New York, 1990.

6. Hoptner, W.; Mehring, M.; Von Schutz, J. U.; Wolf, H. C.; Morra, B. S.; Enkelmann, V.; Wegner, G. *Chem. Phys.* **1982**, *73*, 253–261.

7. Jerome, D.; Mazaud, A.; Ribault, M.; Bechgaard, J. *J. Phys. Lett.* **1980**, *41*, L–95.

8. Parkin, S. S. P.; Engler, E. M.; Schumaker, R. R.; Lagier, R.; Lee, V. Y.; Scott, J. C.; Greene, R. L. *Phys. Rev. Lett.* **1983**, *50*, 270–273.

9. Urayama, H.; Yamochi, H.; Saito, G.; Nozawa, K.; Sugano, T.; Kinoshita, M.; Sato, S.; Oshima, K.; Kawamoto, A.; Tanaka, J. *Chem. Lett.* **1988**, 55.

10. Kini, A. M.; Geiser, U.; Wang, H. H.; Carlson, K. D.; Williams, J. M.; Kwok, W. K.; Vandervoort, K. G.; Thompson, J. E.; Stupka, D. L.; Jung, D.; Whangbo, M.-H. *Inorg. Chem.* **1990**, *29*, 2555–2557.

11. Haddon, R. C.; Glarum, S. H.; Chichester, S. V.; Ramirez, A. P.; Zimmerman, N. M. *Phys. Rev.* **1991**, *43B*, 2642–2647.

12. Haddon, R. C. *Acc. Chem. Res.* **1991**, *21*, 243–249.

13. Kroto, H. W.; Heath, J. R.; O'Brien, S. C.; Curl, R. F.; Smalley, R. E. *Nature (London)* **1985**, *318*, 162–164.

14. Krätschmer, W.; Lamb, L. D.; Fostiropoulos, K.; Huffman, D. R. *Nature* **1990**, *347*, 354–358.

15. Haddon, R. C.; Brus, L. E.; Raghavachari, K. *Chem. Phys. Lett.* **1986**, *125*, 459–464; **1986**, *131*, 165–169.

16. Curl, R. F.; Smalley, R. E. *Science (Washington, D.C.)* **1988**, *242*, 1017–1022.

17. Haufler, R. E.; Conceicao, J.; Chibante, L. P. F.; Chai, Y.; Byrne, N. E.; Flanagan, S.; Haley, M. M.; O'Brien, S. C.; Pan, C.; Xiao, Z.; Billups, W. E.; Ciufolini, M. A.; Hauge, R. H.; Margrave, J. L.; Wilson, L. J.; Curl, R. F.; Smalley, R. E. *J. Phys. Chem.* **1990**, *94*, 8634–8636.

18. Allemand, P.-M.; Koch, A.; Wudl, F.; Rubin, Y.; Diederich, F.; Alvarez, M. M.; Anz, S. J.; Whetten, R. L. *J. Am. Chem. Soc.* **1991**, *113*, 1050–1051.

19. Haddon, R. C.; Hebard, A. F.; Rosseinsky, M. J.; Murphy, D. W.; Duclos, S. J.; Lyons, K. B.; Miller, B.; Rosamilia, J. M.; Fleming, R. M.; Kortan, A. R.; Glarum, S. H.; Makhija, A. V.; Muller, A. J.; Eick, R. H.; Zahurak, S. M.; Tycko, R.; Dabbagh, G.; Thiel, F. A. *Nature (London)* **1991**, *350*, 320–322.

20. Fleming, R. M.; Siegrist, T.; Marsh, P. M.; Hessen, B.; Kortan, A. R.; Murphy, D. W.; Haddon, R. C.; Tycko, R.; Dabbagh, G.; Mujsce, A. M.; Kaplan, M. L.; Zahurak, S. M. *Mater. Res.* in press.

21. Hebard, A. F.; Haddon, R. C.; Fleming, R. M.; Kortan, A. R. *Appl. Phys. Lett.* in press.

22. Whetten, R. L.; Alvarez, M. M.; Anz, S. J.; Schriver, K. E.; Beck, R. D.; Diederich, F. N.; Rubin, Y.; Ettl, R.; Foote, C. S.; Darmanyan, A. P.; Arbogast, J. W. *Mater. Res.* in press.

23. Haufler, R. E.; Wang, L.-S.; Chibante, L. P. F.; Jin, C.; Conceicao, J. J.; Cahi, Y.; Smalley, R. E.; *Chem. Phys. Lett.* **1991**, *179*, 449–454.

24. Miller, B.; Rosamilia, J. M.; Dabbagh, G.; Tycko, R.; Haddon, R. C.; Muller, A. J.; Wilson, W. L.; Murphy, D. W.; Hebard, A. F. *J. Am. Chem. Soc.* **1991**, *113*, 6291–6293.

25. Kochanski, G. P.; Hebard, A. F.; Haddon, R. C; Fiory, A. T. *Science (Washington, D.C.)* submitted, 1991.

26. Hebard, A. F.; Rosseinsky, M. J.; Haddon, R. C.; Murphy, D. W.; Glarum, S. H.; Palstra, T. T. M.; Ramirez, A. P.; Kortan, A. R. *Nature (London)* **1991**, *350*, 600–601.

27. Holczer, K.; Klein, O.; Huang, S.-M.; Kaner, R. B.; Fu, K.-J.; Whetten, R. L.; Diederich, F. *Science (Washington, D.C.)* **1991,** *252,* 1154–1157.

28. Stephens, P. W.; Mihaly, L.; Lee, P. L.; Whetten, R. L.; Huang, S.-M.; Kaner, R.; Diederich, F.; Holczer, K. *Nature (London)* **1991,** *351,* 632–634.

29. Rosseinsky, M. J.; Ramirez, A. P.; Glarum, S. H.; Murphy, D. W.; Haddon, R. C.; Hebard, A. F.; Palstra, T. T. M.; Kortan, A. R.; Zahurak, S. M.; Makhija, A. V. *Phys. Rev. Lett.* **1991,** *66,* 2830–2832.

30. Tycko, R.; Dabbagh, G.; Rosseinsky, M. J.; Murphy, D. W.; Fleming, R. M.; Ramirez, A. P.; Tully, J. C. *Science (Washington, D.C.)* **1991,** *253,* 884–886.

31. Zhou, O.; Fischer, J. E.; Coustel, N.; Kycia, S.; Zhu, Q.; McGhie, A. R.; Romanow, W. J.; McCauley, J. P., Jr.; Smith, A. B., III; Cox, D. E. *Nature (London)* **1991,** *351,* 462–464.

32. Hoffmann, R. *Solids and Surfaces*; VCH: New York, 1988.

33. Newton, M. D.; Stanton, R. E. *J. Am. Chem. Soc.* **1986,** *108,* 2569–2470.

34. Lichtenberger, D. L.; Jatcko, M. E.; Nebesny, K. W.; Ray, C. D.; Huffman, D. R.; Lamb, L. D. *Mater. Res.* in press.

35. Weaver, J. H.; Martins, J. L.; Komeda, T.; Chen, Y.; Ohno, T. R.; Kroll, G. H.; Troullier, N.; Haufler, R. E.; Smalley, R. E. *Phys. Rev. Lett.* **1991,** *66,* 1741–1744.

36. Hoffmann. R. J. *Chem. Phys.* **1963,** *39,* 1397–1412.

37. Cordes, A. W.; Haddon, R. C.; Oakley, R. T.; Schneemeyer, L. F.; Waszczak, J. V.; Young, K. M.; Zimmerman, N. M. *J. Am. Chem. Soc.* **1991,** *113,* 582–588.

38. Eland, J. H. D. *Photoelectron Spectroscopy*; Butterworth: London, 1984; pp 116–121.

39. Whangbo, M.-H.; Willams, J. M.; Leung, P. C. W.; Beno, M. A.; Emge, T. J.; Wang, H. H.; Carlson, K. D.; Crabtree, G. W. *J. Am. Chem. Soc.* **1985,** *107,* 5815–5816.

40. Ramirez, R.; Bohm, M. C. *Int. J. Quant. Chem.* **1988,** *34,* 571–594.

41. Saito, S.; Oshiyama, A. *Phys. Rev. Lett.* **1991,** *66,* 2637–2640.

42. Benning, P. J.; Martins, J. L.; Weaver, J. H.; Chibante, L. B. F.; Smalley, R. E. *Science (Washington, D.C.)* **1991,** *252,* 1417–1419.

43. Wertheim, G. K.; Rowe, J. E.; Buchanan, D. N. E.; Chaban, E. E.; Hebard, A. F.; Kortan, A. R.; Makhija, A. V.; Haddon, R. C. *Science (Washington, D.C.)* **1991,** *2521,* 1419–1421.

44. Tong, W. M.; Ohlberg, D. A. A.; You, H. K.; Williams, R. S.; Anz, S. J.; Alvarez, M. M.; Whetten, R. L.; Rubin, Y.; Diederich, F. N. *J. Phys. Chem.* **1991**, *95*, 4709–4712.

45. Johnson, K. H.; McHenry, M. E.; Clougherty, D. P. *Physica* submitted, 1991.

46. Phillips, J. C., unpublished.

47. Martins, J. L.; Troullier, N.; Schabel, M., unpublished.

48. Chakravarty, S.; Kivelson, S. *Phys. Rev. Lett.* submitted, 1991.

49. Fleming, R. M.; Ramirez, A. P.; Rosseinsky, M. J.; Murphy, D. W.; Haddon, R. C.; Zahurak, S. M.; Makhija, A. V. *Nature (London)* **1991**, *352*, 787–788.

Received August 26, 1991

Chapter 6

Crystal Structure of Osmylated C_{60}

Confirmation of the Soccer-Ball Framework

Joel M. Hawkins, Axel Meyer, Timothy A. Lewis, and Stefan Loren

Department of Chemistry, University of California, Berkeley, CA 94720

An X-ray crystal structure of $C_{60}(OsO_4)$(4-tert-butylpyridine)$_2$ reveals the soccer-ball-shaped carbon framework of buckminsterfullerene (C_{60}) and confirms the originally proposed structure of C_{60}. Osmylation of C_{60} with OsO_4 and pyridine gives high yields of $C_{60}[OsO_4(pyridine)_2]_n$ (n = 1 or 2 depending on the reaction conditions). $C_{60}[OsO_4(pyridine)_2]_2$ is a mixture of five regioisomers; $C_{60}(OsO_4)(pyridine)_2$ forms regioselectively with the O–Os–O unit positioned across the junction of two six-membered rings (a six–six ring fusion), in agreement with theory. Within the C_{60} moiety of $C_{60}(OsO_4)$(4-tert-butylpyridine)$_2$, the tricoordinate carbons lie within a spherical shell with an average radius of 3.512 (3) Å. Within this shell, C–C bond lengths average 1.386 (9) Å for six–six ring fusions and 1.434 (5) Å for six–five ring fusions (junctions of a six- and a five-membered ring). The O-bonded carbons lie outside of this shell and have approximately tetrahedral geometry. The adjacent carbons are the least distorted from planarity. The remaining carbons have approximately equivalent geometries with sums of C–C–C angles averaging 348.0 (3)°. (Averages are reported with the standard error of the mean in parentheses.)

Krätschmer et al.'s discovery that C_{60} could be prepared and isolated in macroscopic quantities (1) initiated a race for chemists and physicists to either confirm or disprove the soccer-ball structure (2) for C_{60}. The infrared (1, 3), Raman (4), ^{13}C NMR (5–9), and photoelectron spectra (10) of C_{60} were each consistent with icosahedral symmetry and collectively highly supportive of the originally proposed structure, but they did not strictly prove the soccer-ball framework or provide atomic positions. For example, the single peak ^{13}C NMR spectrum (5, 6) did not rule out the possibility of coincident peaks or of a fluxional structure. Furthermore, although the truncated icosahedral structure was favored (9), a truncated dodecahedral structure (which also has icosahedral symmetry) could not be ruled out (11). In January 1991, J. Fraser Stoddart wrote, "The chemical world awaits a detailed single-crystal X-ray

0097–6156/92/0481–0091$06.00/0

diffraction analysis of the structure of C_{60} or C_{70}, or more likely perhaps, of a derivative in the first instance" (12). We obtained a detailed single-crystal X-ray diffraction analysis of a C_{60} derivative, $C_{60}(OsO_4)(4\text{-}tert\text{-butylpyridine})_2$, and this was the first instance (13, 14).

We (15) and others (1, 16) attempted to obtain a crystal structure of underivatized C_{60}, but could not determine specific atomic positions because of extensive disorder in the crystals. The ball-like molecules pack in an ordered fashion, but their nearly spherical symmetry promotes orientational disorder. At ambient temperature, they rotate rapidly in the solid state (7, 8). Because X-ray crystallography can provide detailed information only about molecular features that are repeated in an orderly way throughout the crystal, studies of plain (underivatized) C_{60} reveal its round shape, approximate size, and crystal-packing information, but not details of the carbon framework. For example, Figure 1 shows packing models for C_{60} crystals grown from hexanes (15). Two possible packing arrangements fit the Patteron map, but no detailed information within the clusters could be determined.

We reasoned that if C_{60} could be derivatized in a way that broke its apparent spherical symmetry, it might crystallize with orientational order and allow a detailed crystallographic analysis. Specifically, we needed to regioselectively functionalize C_{60} so that the carbon framework would be organized relative to the attached functional group, with this group serving as a handle to keep the carbon clusters uniformly oriented. We explored various reactions known to add functionality to polycyclic aromatic hydrocarbons, including charge-transfer complexation with electron-deficient or polarized arenes, complexation with transition metals, and osmylation. Osmylation was the first process to give discrete stable products, and so we pursued this reaction in detail (13, 17).

Osmylation

Osmium tetroxide is a powerful yet selective oxidant, and the osmylation of polycyclic aromatic hydrocarbons has been known for many years. For example, anthracene adds 2 equivalents of osmium tetroxide in the presence of pyridine (Equation 1) (18). Our initial osmylations of C_{60} also gave 2:1 stoichiometry: treatment of a homogeneous toluene solution of a ~4:1 C_{60}–C_{70} mixture and 2 equivalents of osmium tetroxide at 0 °C with 5 equivalents of pyridine gave a brown precipitate within 1 min. After 12 h at room temperature, the majority of the C_{60} and C_{70} had reacted, according to thin-layer chromatography (TLC). Filtration and washing with toluene gave an 81% yield of osmate ester corresponding to the addition of two osmium(VI) units per carbon cluster (Scheme I) (17).

Initial characterization of the osmate esters was hampered by low solubility, the lack of definitive ^{1}H NMR signals, and weak and complicated ^{13}C NMR signals, but general information could be obtained. The presence of the diolatodioxobis(amine)osmium(VI) ester moiety was established by its characteristic $\nu^{as}(OsO_2)$ IR band at 836 cm^{-1} (18–20), pyridine resonances were visible in the ^{1}H NMR spectrum, and the 2:1 stoichiometry was consistent with

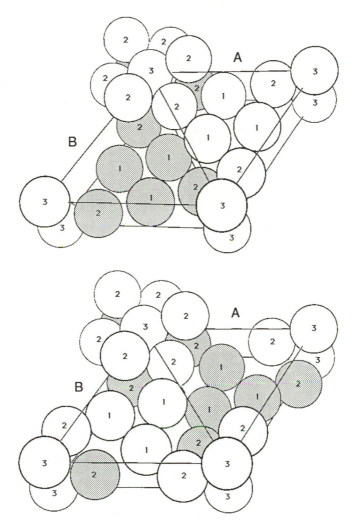

Figure 1. "Triangle" (top) and "triskele" (bottom) models for packing of C_{60} molecules (represented by spheres of radius 3.0 Å) in a crystal grown from hexanes. Molecules in the z = 3/4 plane are shaded. Two symmetry-independent molecules are labeled 1 and 2. A possible 13th molecule in the unit cell (disordered) is labeled 3. (Reproduced with permission from reference 15. Copyright 1991 Royal Society of Chemistry.)

the elemental analysis. The presence of the intact C_{60} skeleton in the osmate esters was established by thermal reversion to C_{60} under vacuum, a type of reaction known for rhenium(V) (21). This reaction was first observed in the electron-impact (EI) mass chromatogram (Figure 2). Heating the osmate ester in the mass spectrometer probe under vacuum at ~290 °C and recording EI spectra at 0.45-min intervals gave spectra of OsO_4 and pyridine (maximum at

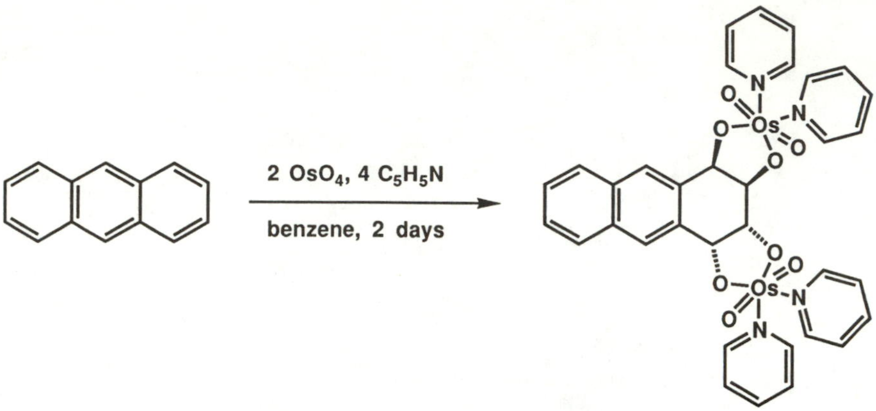

Equation 1.   Osmylation of anthracene.

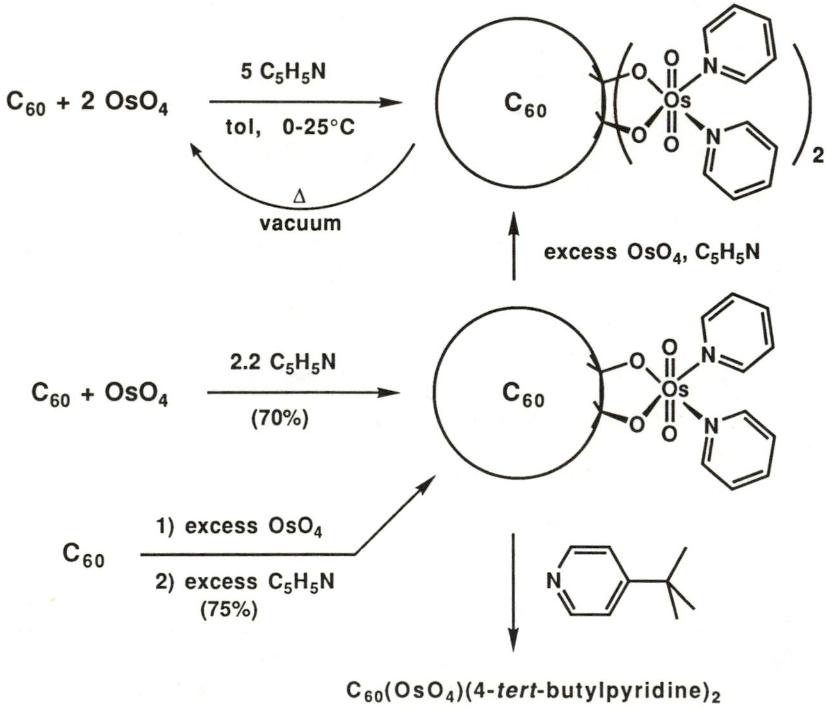

Scheme I.   Osmylation of C_{60} giving 1:1 and 2:1 adducts.

1.8 min) followed by the spectrum of C_{60} (maximum at 5.85 min).   Prolonged heating was required to detect the C_{60} because of its low volatility; this condition was also observed for the spectrum of pure C_{60}.   A preparative version of this experiment whereby the osmate ester was heated under vacuum for 2 min (heat gun, 0.05 mmHg) gave a 47% combined yield of C_{60} and C_{70} (enriched in C_{60} relative to the starting material) as determined by HPLC with respect to

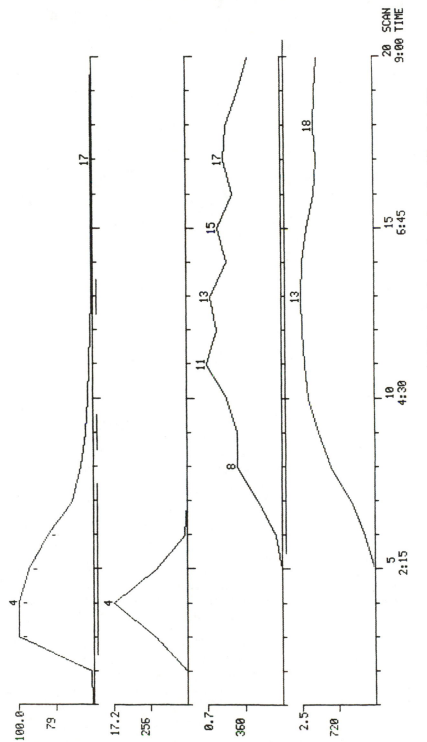

Figure 2. Mass chromatogram of $C_{60}(OsO_4)_2(pyridine)_4$ showing the detection of pyridine (m/z 79), OsO_4 (m/z 256), and C_{60} (m/z 360 and 720) as a function of time (min) at ~ 290 °C.

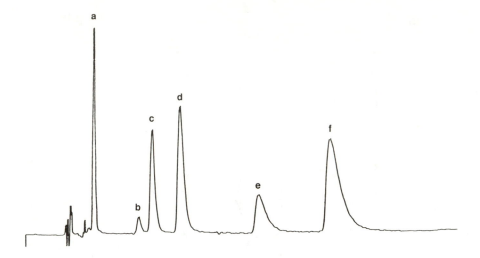

Figure 3. HPLC trace of the crude reaction mixture from the osmylation of C_{60} showing $C_{60}(OsO_4)(pyridine)_2$ (peak a) and regioisomers of $C_{60}(OsO_4)_2(pyridine)_4$ (peaks b–f).

a naphthalene standard. The osmate ester was free of unreacted C_{60} according to IR spectroscopy and TLC, so the C_{60} must have been re-formed upon heating. These experiments were the first to establish that heteroatom functionality can be added to C_{60} without disrupting its carbon framework (*17*).

The initial osmylation conditions were developed to optimize the yield of the precipitated osmate ester. Because this precipitate has 2:1 stoichiometry, these conditions favored the addition of two osmyl units to C_{60}. The 1:1 adduct was more desirable, however, as it has only two possible regioisomers, in contrast to the bisosmylated material, which has 54 possible regioisomers. Chromatographic analysis of the crude reaction mixture from the osmylation of pure C_{60} revealed six peaks: five peaks corresponded to the precipitate that collectively analyzes with 2:1 stoichiometry, and a single sharp peak corresponded to toluene-soluble material (Figure 3). Use of 1 equivalent of OsO_4 increased the yield of the toluene-soluble material to 70% (Scheme I). Osmylation in the absence of pyridine, followed by dimer disruption with pyridine (*20*), gave the same species in 75% yield. The toluene-soluble material was shown to have 1:1 stoichiometry by converting it to the mixture of 2:1 adducts upon further exposure to the osmylation conditions. Solubility and crystal quality were improved by exchanging the pyridine ligands for 4-*tert*-butylpyridine (Scheme I).

The observation of a single sharp chromatographic peak for the 1:1 adduct suggested that it is a single regioisomer, rather than a mixture of the two regioisomers that are possible from the soccer-ball structure for C_{60}. This single regioisomer would be found if the bisoxygenation was strongly favored across one of the two unique bonds in C_{60}, the junction of two six-membered

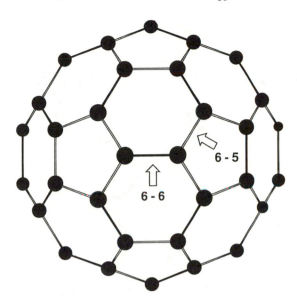

Figure 4. Two unique bonds in C_{60}, the junction of two six-membered rings and the junction of a six- and a five-membered ring.

rings, or the junction of a six- and a five-membered ring (Figure 4). Regioselective osmylation would fix the position of the C_{60} carbon framework relative to the osmyl unit, with the osmyl unit breaking the pseudospherical symmetry of C_{60} as required for an ordered crystal. The 1:1 adduct, $C_{60}(OsO_4)(4\text{-}tert\text{-butylpyridine})_2$, indeed gave a sufficiently ordered crystal for the determination of atomic positions by X-ray crystallographic analysis (*13*).

Analysis of Crystal Structure

The crystal structure of $C_{60}(OsO_4)(4\text{-}tert\text{-butylpyridine})_2$ proves the soccer-ball-shaped arrangement of carbon atoms in C_{60} by clearly showing the 32 faces of the carbon cluster composed of 20 six-membered rings fused with 12 five-membered rings (Figures 5 and 6). No two five-membered rings are fused together, and each six-membered ring is fused to alternating six- and five-membered rings. Sixteen equivalent molecules of $C_{60}(OsO_4)(4\text{-}tert\text{-butylpyr-idine})_2$ occur in the unit cell, which also contains 2.5 toluenes of crystallization per osmate ester (Figure 7). A typical close contact is shown in Figure 8. Intermolecular contacts between C_{60} moieties include carbon–carbon distances as small as 3.29 (4) Å.

The crystal structure shows that the O–Os–O unit has added across the fusion of two six-membered rings. The ^{13}C NMR spectrum of the 1:1 adduct recorded before crystallization indicates that the isomer in the crystal is the only one present within the limits of detection (*22*). The osmylation appears

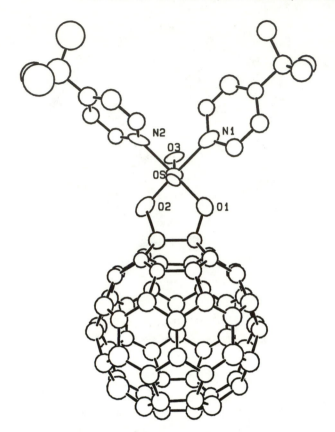

Figure 5. ORTEP drawing (50% ellipsoids) of the 1:1 C_{60}–osmium tetroxide adduct $C_{60}(OsO_4)$(4-tert-butylpyridine)$_2$ showing the relationship of the osmyl unit with the carbon cluster. [$C_{60}(OsO_4)$(4-tert-butylpyridine)$_2$· 2.5 toluene]: tetragonal, space group I4$_1$/a, a = 30.751 (5) Å, c = 24.800 (7) Å, V = 23452 (14) Å^3, Z = 16. All atoms were located and all positions were refined. Osmium, oxygen, and nitrogen were anisotropic, and all others were isotropic thermal parameters. R = 10.6%, wR = 10.3%, GOF = 1.77. 442 parameters, 3668 observed data. (Reproduced with permission from reference 13. Copyright 1991 American Association for the Advancement of Science.)

to be quite regioselective, and the favored approach of osmium tetroxide agrees with the regiochemistry predicted by Hückel calculations on C_{60} and the principle of least motion or minimum electronic reorganization (Dias, J. R., personal communication) *(23, 24)*. Extended Hückel calculations on a Tektronix CAChe system qualitatively agree with these HMO calculations. Chemical reactivity, especially regioselectivity, can thus serve as a probe of fullerene structure and bonding in a way that complements theoretical and spectroscopic techniques.

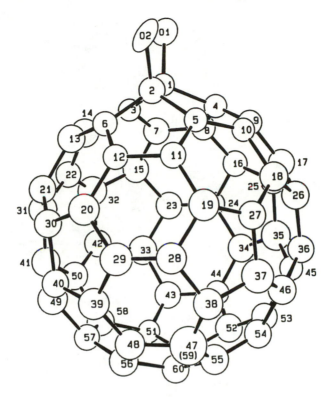

Figure 6. ORTEP drawing (50% ellipsoids) of the 1:1 C$_{60}$–osmium tetroxide adduct C$_{60}$(OsO$_4$)(4-tert-butylpyridine)$_2$ showing the geometry of the C$_{60}$O$_2$ unit and the numbering scheme. (Reproduced with permission from reference 13. Copyright 1991 American Association for the Advancement of Science.)

Analysis of the structure of the carbon cluster was aided by its apparent symmetry. A histogram of distances from the calculated center of the cluster to each of the carbon atoms in the cluster shows two groupings (Figure 9). The tricoordinate carbons C-3–C-60 all lie within a spherical shell of radius 3.46–3.56 Å, with an average distance of 3.512 (3) Å from the center of the cluster. (Averages are reported with the standard error of the mean in parentheses.) The tetracoordinate oxygen-bonded carbons C-1 and C-2 lie significantly outside of this shell at distances of 3.80 (2) and 3.81 (3) Å from the center, respectively. They have approximately tetrahedral geometry with sums of C–C–C angles equal to 330°, slightly more than 328° (the sum for an ideal tetrahedral atom). The proximate carbons, C-3–C-6, are the least distorted from planarity within the cluster, with sums of C–C–C angles averaging 353 (1)°, compared with 360° for a planar atom. The remaining carbons, C-7–C-60, have approximately equivalent geometries with sums of C–C–C angles ranging from 344° to 351° (Figure 10). The average sum, 348.0 (3)°, equals the

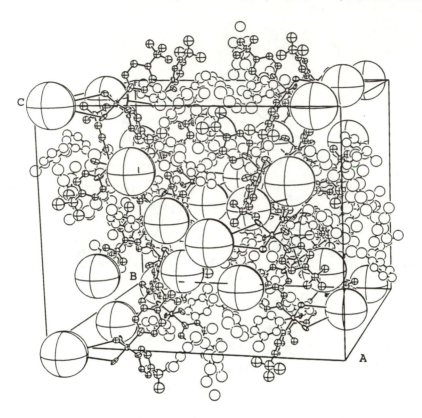

Figure 7. Unit cell for [C_{60}(OsO_4)(4-tert-butylpyridine)_2· 2.5 toluene] showing the 16 equivalent molecules of C_{60}(OsO_4)(4-tert-butylpyridine)_2. The C_{60} moieties are represented by undersized spheres to show the packing arrangement more clearly.

value for an ideal junction of two regular hexagons and a regular pentagon. All 60 carbons within the cluster are pyramidalized concave inwards.

The C-1–C-3, C-1–C-4, C-2–C-5, and C-2–C-6 bond lengths average 1.53 (3) Å, comparable with normal $C(sp^3)–C(sp^2)$ single bonds. The geometry of the OsO_2N_2(diolate) unit is similar to that observed for conventional arene adducts, although our C-1–C-2 bond length (1.62 (4) Å) appears to be longer than the corresponding bonds in the other structures (1.40 (4) to 1.54 (2) Å) (*18*).

The five- and six-membered carbocyclic rings not containing C-1 and C-2 are planar with deviations from least-squares planes less than 0.05 (3) Å. In contrast, tetracoordinate carbons C-1 and C-2 lie 0.22 (2) to 0.30 (3) Å outside of the planes defined by the other carbons in the rings that contain them. Excluding bonds to C-1 and C-2, the average C–C bond lengths are 1.386 (9) Å for six–six ring fusions (junctions of two six-membered rings), and 1.434 (5) Å for six–five ring fusions (junctions of a six- and a five-membered ring).

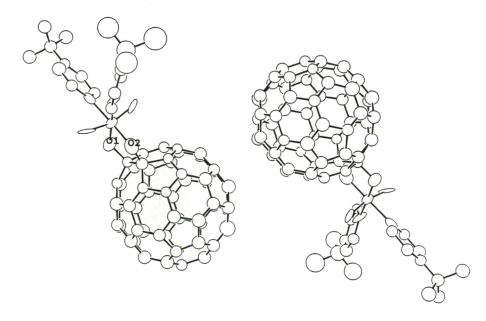

Figure 8. Typical close contact between C$_{60}$(OsO$_4$)(4-tert-butylpyridine)$_2$ *molecules in the solid state.*

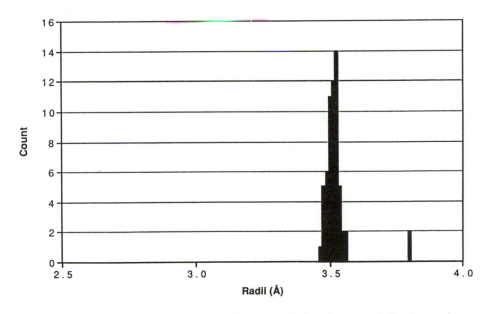

Figure 9. Histogram of distances from the calculated center of the C$_{60}$ *moiety to* C-1–C-60 *in* C$_{60}$(OsO$_4$)(4-tert-butylpyridine)$_2$.

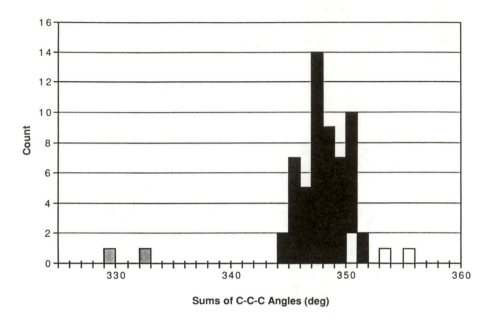

Figure 10. Histogram of sums of C–C–C angles in $C_{60}(OsO_4)$(4-tert-butyl-pyridine)$_2$: C-1–C-2 (shaded), C-3–C-6 (white), and C-7–C-60 (black).

Histograms for these two types of bond lengths show overlap between the two sets, but the two averages are statistically different (Figure 11). [Carbon–carbon coupling constants ($^1J_{CC}$) reveal a more distinct division between the six–six and six–five ring fusions in $C_{60}(OsO_4)$(4-*tert*-butylpyridine)$_2$ (22).] These two average bond lengths are within the range of values predicted by theory for the two types of bonds in C_{60} (1.36 to 1.42 Å for six–six ring fusions, and 0.02 to 0.08 Å longer for six–five ring fusions) (25), and within experimental error of the two (unassigned) bond lengths in C_{60} determined from $^{13}C–^{13}C$ magnetic dipolar coupling (1.40 ± 0.015 and 1.45 ± 0.015 Å) (9).

Summary

Osmylation broke the pseudospherical symmetry of C_{60} and allowed the formation of an ordered crystal and structural analysis of the carbon framework for the first time. Osmylation offered the advantages of rapid but selective addition to the cluster, variable ligands on osmium for tuning solubility and crystal quality, and the addition of a symmetrical O–Os–O species to minimize the number of possible isomers. The carbon cluster in $C_{60}(OsO_4)$(4-*tert*-butyl-pyridine)$_2$ appears to be fairly undistorted from an ideal soccer-ball shape for C-7 through C-60, and highly perturbed in the local environment around the added oxygens. The undistorted region serves as a model for C_{60} and provides information that underivatized C_{60} cannot give (22). The added function-

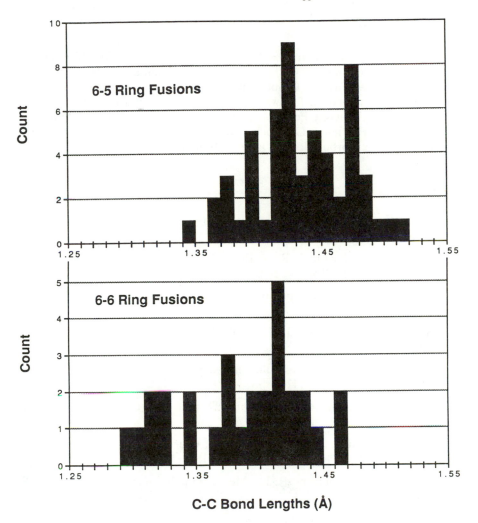

Figure 11. Histograms of carbon–carbon bond lengths for C-3–C-60 in C$_{60}$(OsO$_4$)(4-tert-butylpyridine)$_2$ showing six–five ring fusions (top) and six–six ring fusions (bottom).

al group allows tailoring properties in order to go beyond buckminsterfullerene in the pursuit of new and unusual molecules. Specifically, the novel cup-and-band shaped conjugated π-systems of the 1:1 and 2:1 C$_{60}$–OsO$_4$ adducts should be interesting from chemical, spectroscopic, and theoretical perspectives.

Acknowledgments

J. M. Hawkins is grateful to the National Science Foundation (Presidential Young Investigator Award, CHE–8857453), the Camille and Henry Dreyfus Foundation (New Faculty Grant), the Merck Sharp & Dohme Research Laboratories (postdoctoral fellowship for A. Meyer), the Shell Oil Company Foundation (Shell Faculty Fellowship), the Xerox Corporation, the Monsanto Company, and Hoffmann-La Roche Inc. for financial support. S. Loren thanks the Syntex Corporation for a graduate fellowship.

References

1. Krätschmer, W.; Lamb, L. D.; Fostiropoulos, K.; Huffman, D. R. *Nature (London)* **1990,** *347,* 354.

2. Kroto, H. W.; Heath, J. R.; O'Brien, S. C.; Curl, R. F.; Smalley, R. E. *Nature (London)* **1985,** *318,* 162.

3. Krätschmer, W.; Fostiropoulos, K.; Huffman, D. R. *Chem. Phys. Lett.* **1990,** *170,* 167.

4. Bethune, D. S.; Meijer, G.; Tang, W. C.; Rosen, H. J. *Chem. Phys. Lett.* **1990,** *174,* 219.

5. Taylor, R.; Hare, J. P.; Abdul-Sada, A. K.; Kroto, H. W. *J. Chem. Soc. Chem. Commun.* **1990,** 1423.

6. Johnson, R. D.; Meijer, G.; Bethune, D. S. *J. Am. Chem. Soc.* **1990,** *112,* 8983.

7. Tycko, R.; Haddon, R. C.; Dabbagh, G.; Glarum, S. H.; Douglass, D. C.; Mujsce, A. M. *J. Phys. Chem.* **1991,** *95,* 518.

8. Yannoni, C. S.; Johnson, R. D.; Meijer, G.; Bethune, D. S.; Salem, J. R. *J. Phys. Chem.* **1991,** *95,* 9.

9. Yannoni, C. S.; Bernier, P. P.; Bethune, D. S.; Meijer, G.; Salem, J. R. *J. Am. Chem. Soc.* **1991,** *113,* 3190.

10. Lichtenberger, D. L.; Nebesny, K. W.; Ray, C. D.; Huffman, D. R.; Lamb, L. D. *Chem. Phys. Lett.* **1991,** *176,* 203.

11. Shibuya, T.; Yoshitani, M. *Chem. Phys. Lett.* **1987,** *137,* 13.

12. Stoddart, J. F. *Angew. Chem. Int. Ed. Engl.* **1991,** *30,* 70.

13. Hawkins, J. M.; Meyer, A.; Lewis, T. A.; Loren, S.; Hollander, F. J. *Science (Washington, D.C.)* **1991,** *252,* 312.

14. Fagan subsequently reported the crystal structure of the platinum complex $[(C_6H_5)_3P]_2Pt(\eta^2\text{-}C_{60})$. Fagan, P. J.; Calabrese, J. C.; Malone, B. *Science (Washington, D.C.)* **1991,** *252,* 1160.

15. Hawkins, J. M.; Lewis, T. A.; Loren, S. D.; Meyer, A.; Heath, J. R.; Saykally, R. J.; Hollander, F. J. *J. Chem. Soc. Chem. Commun.* **1991**, 775.

16. Fleming, R. M.; Siegrist, T.; Marsh, P. M; Hessen, B.; Kortan, A. R.; Murphy, D. W.; Haddon, R. C.; Tycko, R.; Dabbagh, G.; Mujsce, A. M.; Kaplan, M. L.; Zahurak, S. M. *Mater. Res. Soc. Symp. Proc.* **1991**, *206*, 691.

17. Hawkins, J. M.; Lewis, T. A.; Loren, S. D.; Meyer, A.; Heath, J. R.; Shibato, Y.; Saykally, R. J. *J. Org. Chem.* **1990**, *55*, 6250.

18. Wallis, J. M.; Kochi, J. K. *J. Am. Chem. Soc.* **1988**, *110*, 8207.

19. Schroder, M. *Chem. Rev.* **1980**, *80*, 187.

20. Collin, R. J.; Jones, J.; Griffith, W. P. *J. Chem. Soc. Dalton Trans.* **1974**, 1094.

21. Pearlstein, R. M.; Davison, A. *Polyhedron* **1988**, *7*, 1981.

22. Hawkins, J. M.; Loren, S.; Meyer, A.; Nunlist, R. *J. Am. Chem. Soc.* **1991**, *113*, 7770.

23. Dias, J. R. *J. Chem. Ed.* **1989**, *66*, 1012.

24. Amic, D.; Trinajstic, N. *J. Chem. Soc. Perkin Trans. 2* **1990**, 1595.

25. Weltner, W., Jr.; Van Zee, R. *J. Chem. Rev.* **1989**, *89*, 1713.

Received September 6, 1991

Chapter 7

One- and Two-Dimensional NMR Studies

C_{60} and C_{70} in Solution and in the Solid State

Robert D. Johnson, Costantino S. Yannoni, Jesse R. Salem, Gerard Meijer[1], and Donald S. Bethune

IBM Research Division, Almaden Research Center, San Jose, CA 95120-6099

We have investigated the structure and dynamics of C_{60} and C_{70} with ^{13}C NMR spectroscopy in both the liquid and the solid state. The spectra, bonding topology, and resonance assignments obtained using solution NMR techniques strongly support the truncated icosahedral soccer-ball structure for C_{60} and the D_{5h} rugby-ball structure for C_{70}. Solid-state NMR spectroscopy reveals that C_{60} rotates rapidly at 295 K. Cooling to 77 K leads to a powder pattern yielding the chemical-shift tensor. The C_{60} bond lengths are determined from measurements of the dipolar coupling of adjacent ^{13}C atoms in isotopically enriched samples.

The realization of geodesic structures in chemistry occurred in experiments on carbon clusters. In 1985 Smalley, Kroto, and co-workers ascribed a closed-shell, "soccer ball" structure to an observed 60-atom pure carbon cluster of unusual prominence (1), and the great stability of this structure was established 5 years later by the production of bulk quantities of fullerenes (2–4). The proposed structures (1, 5) for C_{60} and C_{70} shown in Figures 1 and 2 are carbon shells consisting of hexagons and pentagons; the 12 pentagons in each structure produce torsional strain, creating curvature that leads to closure.

The structure, dynamics, and electronic properties of fullerenes are of great interest to many fields of chemistry, and characterization of these properties is important to current efforts in their derivatization and possible application. NMR spectroscopy is a powerful tool for characterization of both structure and dynamics (6, 7), and is well suited to the study of these all-carbon molecules. We have performed a wide range of NMR experiments on the fullerenes C_{60} and C_{70} to learn more about their symmetries, chemical structures, bonding, and dynamics, and we report results in both the liquid and solid state.

[1]Permanent address: Department of Molecular and Laser Physics, University of Nijmegen, Toernooiveld, 6525 ED Nijmegen, Netherlands

0097–6156/92/0481–0107$06.00/0

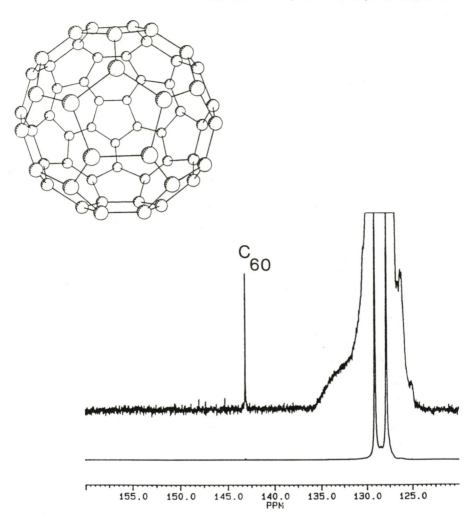

Figure 1. ^{13}C NMR spectrum of 80 μg of C_{60}–C_{70} in benzene-$^2H_6$. The spectrum was obtained on a Bruker AM500 NMR spectrometer operating at 125 MHz at 25 °C from 49,000 scans.

Liquid-State NMR Results

The ^{13}C NMR spectrum of C_{60} is shown in Figure 1 (8, 9). The sample is small by NMR standards; ~80 μg of a C_{60}–C_{70} mixture was obtained by sublimation from graphitic soot onto quartz slides and was subsequently washed in

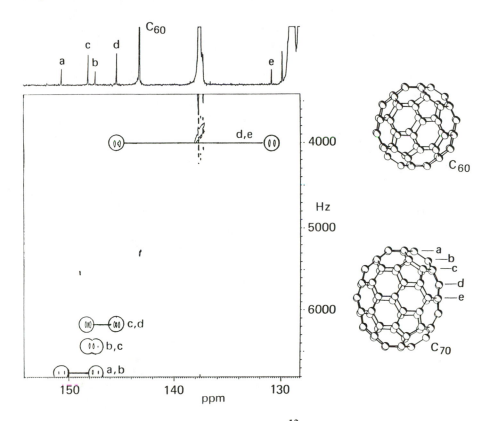

Figure 2. Upper trace: One-dimensional (1D) ^{13}C NMR spectrum of 60 μg of ~10× ^{13}C-enriched C_{70} with C_{60} in toluene-$^2H_8$. Matrix: two-dimensional (2D) NMR INADEQUATE spectrum of this sample. Doublets (circled) occur at a common double-quantum frequency for two resonances of bonded carbons, allowing the bonding connectivity to be made. The spectrum was obtained in the presence of $Cr(ac)_3$, using the pulse sequence of Mareci and Freeman with a refocusing delay of 5 ms, a sweep-width of 13 kHz digitized into 8 kword data sets for horizontal dimension, and a sweep-width of 7 kHz for the double-quantum dimension digitized into 512 blocks, with 512 scans per block.

benzene. This sample was then concentrated to 100 μL, and $Cr(ac)_3$ was added to allow a practical pulse repetition time for the NMR experiment. The mass spectrum of the sample showed a 4:1 ratio of C_{60} to C_{70}. The ^{13}C NMR spectrum was obtained without ^1H decoupling, and shows an intense ^1H-coupled solvent resonance due to benzene, a resonance at 142.8 ppm that we assign to C_{60}, and several smaller lines at 150.6, 148.0, 147.3, and 145.3 ppm, whose net intensity are consistent with the mass spectrum (the fifth line of C_{70} is under the solvent resonance at 130.8 ppm; *see* Figure 2 and ref. 8). The C_{60} ^{13}C NMR spectrum is remarkable for its simplicity. The single resonance observed indicates that all 60 carbons are chemically equivalent on the time scale of the NMR experiment and show no coupling to protons. The position of this resonance at 142.8 ppm is in the region characteristic of aromatic or olefinic carbons possessing some torsional strain, and is consistent with the shifts for similar carbons (6) in azulene (140.2 ppm), fluorene (141.6 and 143.2 ppm), and 3,5,8-trimethylaceheptylene (146.8 ppm), which are polycyclic hydrocarbons possessing some ring strain. The ^{13}C NMR spectrum of C_{60} thus strongly supports the proposed soccer-ball geometry shown in Figure 1.

The upper trace in Figure 2 shows the ^{13}C NMR spectrum of a mixture of ^{13}C-enriched C_{60} and C_{70} in toluene. The $\sim10\times$ enrichment was achieved by using cored carbon rods loaded with amorphous 98% ^{13}C powder (Cambridge Isotopes) in an arc fullerene generator (10, 11). The spectrum shows the C_{60} resonance, and the five C_{70} resonances with intensity ratios 10:20:10:20:10, which were interpreted (8) to support the proposed C_{70} structure with D_{5h} symmetry (5) shown in Figure 2. The C_{70} structure is similar to that of C_{60}, but with the insertion of 10 carbons around the equator, and with a 36° relative hemispheric rotation. The five ^{13}C resonances in C_{70} are well separated, and, when adjacent carbons are both ^{13}C isotopes, will split into doublets because of $^1J_{CC}$ coupling. The C_{70} resonances are predominantly singlet in nature. As our enrichment is about 10×, 90% of the ^{13}C nuclei in the sample arise from the 98% ^{13}C amorphous carbon insert. The coupling pattern of these resonances, which show doublets of 5% intensity from $^{13}C-^{13}C$ pairing, indicate that the ^{13}C atoms in C_{70} are randomly mixed with the ^{12}C atoms. If ^{13}C atoms from the enriched core of the graphite rod were consistently incorporated in groups or pairs, the NMR resonances would all be doublets because of one-bond coupling of adjacent ^{13}C atoms. That they are not shows that **the fullerenes form by condensation of a gas where the carbon atoms are mixed on an atomic scale,** despite the initial spatial separation of the ^{12}C (natural abundance 1% ^{13}C) and ^{13}C graphites.

The NMR spectra of fullerenes with lower symmetry than C_{60} reveal much structural information. The C_{70} resonance at 130.8 ppm is far upfield of the other fullerene peaks, and, as noted by Taylor et al. (8), although four of the C_{70} carbons are on five- and six-membered rings, the belt carbon (e in Figure 2) in C_{70} is unique in that it belongs only to six-membered rings. Its chemical shift should then be similar to that of benzo[a]pyrene (125.5 and 123.8 ppm) (12), and was assigned to the resonance at 130.8 ppm (8); the other four resonances were assigned using insights on torsional strain. The D_{5h} structure of C_{70} shows a linear connectivity of the five carbons with differing intensities. We have been able to map this bonding topology by using the two-dimensional (2D) NMR "INADEQUATE" experiment (13), which correlates the ^{13}C NMR line of a carbon to that of its bonded neighbor (14–16).

The basis of the INADEQUATE experiment is the scalar J coupling of adjacent ^{13}C atoms, which enables the excitation and detection of a shared double-quantum coherence. The 2D spectrum then allows construction of a bonding connectivity map of the molecule. A ^{13}C-enriched C_{70} sample was used to increase the occurrence of adjacent ^{13}C carbon atoms. The 2D INADEQUATE spectrum is shown in Figure 2. The upper trace is the one-dimensional (1D) spectrum. The horizontal axis is the chemical shift, and the vertical axis represents the double-quantum frequency dimension. Two bonded carbons share a double-quantum frequency; peaks in the 2D spectrum occur at their individual chemical shifts along the horizontal dimension and their shared double-quantum frequency in the vertical dimension; this feature allows the correlation to be made. The 2D spectrum shows all four of the bonding connectivities, yielding a linear string of resonances with intensities 10:10:20:20:10, in accord with the structure for C_{70} shown in Figure 2. This topology is asymmetric, so the five C_{70} lines may be **experimentally** assigned. The assignments are shown in Figure 2 and are in accord with those proposed by Taylor et al. (8). The experiment also furnishes the four $^1J_{CC}$ values; $^1J_{a,b}$ = 68 Hz, and $^1J_{d,e}$ = 62 Hz; these results indicate these bonds fusing six-membered rings have substantial π bond order and are more olefinic than aromatic in character (13).

Solid-State NMR Results

Solid-state NMR studies reveal additional aspects of the behavior and properties of C_{60} (17, 18). The nearly spherical shape of the molecules suggests that they may reorient freely in the solid state. Solid-state NMR spectroscopy allows this possibility to be directly checked. Figure 3 shows ^{13}C NMR spectra of solid C_{60} (with a minor amount of C_{70}) obtained at temperatures of 295, 123, 100, and 77 K at 1.4 T (17). The spectrum at 295 K consists of a narrow single peak (70 Hz FWHM) with a chemical shift of 143 ppm, close to the value found for C_{60} in solution (8, 9). As the temperature is lowered, the sharp peak decreases, while a broad asymmetric peak grows in, roughly centered on the sharp peak. The spectrum at 77 K is a powder pattern characteristic of a carbon in a rigid solid. The pattern results from the variation of the magnetic shielding of the ^{13}C nuclei with molecular orientation (19, 20). The components of the chemical-shift tensor describing this anisotropy are found to be 220, 186, and 25 ppm, consistent with expectations for an aromatic molecule (11, 19, 20). The collapse of the line as the sample is warmed to 295 K proves that the molecule reorients on a time scale of $\simeq 1$ ms. At intermediate temperatures the spectrum appears to be a superposition of a narrow line and a broadened line, a finding suggesting the possibility that two phases with different barriers to reorientation may coexist (21), or that there may be a distribution of barriers to rotation in the sample (22).

Spin relaxation mechanisms that arise from molecular motion provide a means to characterize motional rates. For C_{60} in the solid state, a major source of ^{13}C NMR relaxation arises from reorientation, via molecular rotation, of the chemical-shift tensor mentioned. The longitudinal spin relaxation rate of C_{60} was measured for temperatures from 170 to 300 K, and a minimum in this relaxation rate was found at 233 K. This minimum indicates that the rate of molecular reorientation at this temperature is similar to the ^{13}C NMR frequency, which at a magnetic field strength of 7.0 T indicates a reorientational correlation time of ~2 ns at 230 K.

The presence of the rapid molecular reorientation observed with solid-state NMR spectroscopy may pose an obstacle to the experimental determination of the internal atomic structure of these species by X-ray diffraction (2) or

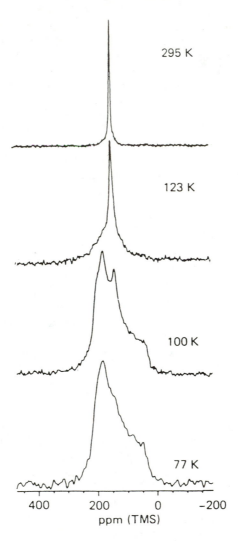

Figure 3. Variable-temperature ^{13}C NMR spectra of solid C_{60}. The sample is ~1 g of C_{60} and C_{70} in a ratio of 10:1.

STM (scanning tunneling microscopy) imaging (23, 24). These techniques have so far yielded only approximate measurements of the center-to-center distances of C_{60}, yielding values of 10–11 Å. On the other hand, solid-state NMR itself offers the possibility of directly determining the bond lengths in the C_{60} molecule by measuring the line splitting due to dipolar coupling of adjacent ^{13}C nuclei (25). This method for measuring bond lengths in orientationally disordered materials was recently quantified (26). To increase the probability of finding two adjacent ^{13}Cs, the clusters were enriched to 6% ^{13}C using the

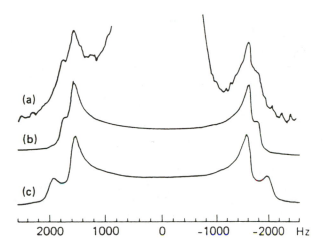

2000   1000   0   -1000   -2000   Hz

Figure 4. (a) ^{13}C *NMR spectrum of* C_{60} *(6%* ^{13}C*) obtained using a Carr–Purcell Meiboom–Gill sequence. The center line has been cropped.* (b) *Simulation of Pake doublets for C–C bond lengths of 1.45 and 1.40 Å, with twice as many long bonds.* (c) *Simulation as in part b, with bond lengths 1.451 and 1.345 Å.*

same method described earlier. The sample was cooled to 77 K to slow the molecular rotation sufficiently to avoid averaging the dipole coupling to zero (*27*). The Carr–Purcell pulse sequence (*28*) was used, with the Meiboom–Gill modification (*29*), to eliminate the effects of chemical shift while retaining the dipolar coupling between ^{13}C spins. In amorphous or polycrystalline samples, this coupling leads to a powder pattern that exhibits a set of relatively sharp Pake doublets (*30*). The splitting of a given doublet depends on the inverse third power of the corresponding carbon–carbon separation.

Figure 4a shows the spectrum obtained using the enriched sample (~6:1 C_{60}–C_{70} ratio) in a field of 1.4 T. Bracketing the central peak due to isolated ^{13}C atoms is a pair of doublets with splittings of 3158 and 3596 Hz. Figure 4b shows a best-fit simulation of this spectrum. Two bond lengths, 1.45 and 1.40 ($\pm$ 0.015) Å, are obtained in satisfactory agreement with the bond lengths calculated for the truncated icosahedral C_{60} structure (*31*), yielding a cage diameter of 7.1 ($\pm$0.07) Å. For comparison, the simulated spectrum for the alternative icosahedrally symmetric C_{60} cluster [the truncated dodecahedron with calculated bond lengths 1.451 and 1.345 Å (*32*)] is shown in Figure 4c. These comparisons lend support to conclusion that C_{60} has the truncated icosahedral (soccer-ball) structure.

Summary and Conclusions

The experiments described here demonstrate that ^{13}C NMR spectroscopy is a powerful technique for investigating many properties of the highly symmetrical hollow carbon molecules with 60 and 70 atoms. In solution, high-resolution NMR spectra reveal that C_{60} has a single resonance at 143 ppm, indicating a strained, olefinic, or aromatic system with high symmetry. This single resonance is strong evidence for a C_{60} soccer-ball geometry. The random distribution of ^{13}C-labeled atoms in 10% enriched C_{70} shows that the fullerenes form by condensation of an atomic gas. A two-dimensional NMR INADEQUATE experiment on ^{13}C-enriched C_{70} reveals the bonding connectivity to be a linear string of the five resonances, with a topology of 10:20:10:20:10, in firm support of the proposed rugby-ball structure with D_{5h} symmetry. In addition, the 2D experiment furnish the four one-bond coupling constants, and the asymmetric bonding topology allows an experimental resonance assignment. Solid-state NMR spectra of C_{60} at ambient temperatures yield a narrow resonance, indicative of rapid isotropic molecular reorientation. Variable-temperature T_1 measurements show that the reorientational correlation time is 2 ns at 230 K. At 77 K, this reorientational time is slowed to more than 0.3 ms, and the resultant ^{13}C NMR spectrum of C_{60} has a powder pattern due to chemical-shift anisotropy, from which the chemical shift tensor components 220, 186, and 25 ppm are obtained. At intermediate temperatures, a narrow peak is superimposed on the powder pattern, a result suggesting a distribution of barriers to molecular motion in the sample, or the presence of an additional phase in the solid state. A Carr–Purcell dipolar experiment on C_{60} in the solid state allows precise determination of the C_{60} bond lengths: 1.45 and 1.40 (±0.015) Å.

Acknowledgments

We gratefully acknowledge stimulating discussions with F. A. L. Anet, G. M. Wallraff, C. G. Wade, and P. P. Bernier. We thank R. D. Kendrick and G. R. May for technical assistance.

References

1. Kroto, H. W.; Heath, J. R.; O'Brien, S. C.; Curl, R. F.; Smalley, R. E. *Nature (London)* **1985**, *318*, 162–163.

2. Krätschmer, W.; Fostiropoulos, K.; Huffman, D. R. *Chem. Phys. Lett.* **1990**, *170*, 167–170.

3. Krätschmer, W.; Lamb, L. D.; Fostiropoulos, K.; Huffman, D. R. *Nature (London)* **1990**, *347*, 351–358.

4. Meijer, G.; Bethune, D. S. *J. Chem. Phys.* **1990**, *93*, 7800.

5. Heath, J. R.; O'Brien, S. C.; Zhang, Q.; Liu, Y.; Curl, R. F.; Kroto, H. W.; Zhang, Q.; Tittel, F. K.; Smalley, R. E. *J. Am. Chem. Soc.* **1985**, *107*, 7779.

6. Kalinowski, H.; Berger, S.; Braun, S. *Carbon-13 NMR Spectroscopy*; Wiley: New York, 1988.

7. Fyfe, C. *Solid State NMR for Chemists*; CFC Press: Guelph, Ontario, Canada, 1983.

8. Taylor, R.; Hare, J. P.; Abdul-Sada, A. K.; Kroto, H. W. *J. Chem. Soc. Chem. Commun.* **1990**, *20*, 1423.

9. Johnson, R. D.; Meijer, G.; Bethune, D. S. *J. Am. Chem. Soc.* **1990**, *112*, 8983.

10. Krätschmer, W.; Sorg, N.; Huffman, D. R. *Surf. Sci.* **1985**, *156*, 814.

11. Haufler, R. E.; Conceicao, J.; Chibante, L. P. F.; Chai, Y.; Byrne, N. E.; Flanagan, S.; Haley, M. M.; O'Brien, S. C.; Pan, C.; Xiao, Z.; Billups, W. E.; Ciufolini, M. A.; Hauge, R. H.; Margrave, J. L.; Wilson, L. J.; Curl, R. F.; Smalley, R. E. *J. Phys. Chem.* **1990**, *94*, 8634.

12. Unkefer, C. J.; London, R. E.; Waley, T. W.; Daub, G. H. *J. Am. Chem. Soc.* **1983**, *105*, 733.

13. Johnson, R. D.; Meijer, G.; Salem, J. R.; Bethune, D. S. *J. Am. Chem. Soc.* **1991**, *113*, 3619–3621.

14. Bax, A.; Freeman, R.; Kempsell, S. P. *J. Am. Chem. Soc.* **1980**, *102*, 4849.

15. Bax, A.; Freeman, R.; Frenkiel, T. A.; Levitt, M. H. *J. Magn. Reson.* **1981**, *43*, 478.

16. Mareci, T. H.; Freeman, R. *J. Magn. Reson.* **1982**, *48*, 158–163.

17. Yannoni, C. S.; Johnson, R. D.; Meijer, G.; Bethune, D. S.; Salem, J. R. *J. Phys. Chem.* **1991**, *95*, 9–10.

18. Tycko, R.; Haddon, R. C.; Dabbagh, F.; Glarum, S. H.; Douglass, D. C.; Mujsce, A. M. *J. Phys. Chem.* **1991**, *95*, 518.

19. Veeman, W. S. *Progress in Nuclear Magnetic Resonance Spectroscopy*; Pergamon: Oxford, England, 1984; pp 193–237.

20. Mehring, M. *High Resolution NMR in Solids*, 2nd ed.; Springer Verlag: Berlin, Germany, 1983, Chapter 7.

21. Shiau, W.-I.; Duesler, E. N.; Paul, I. C.; Curtin, D. Y.; Blann, W. G.; Fyfe, C. A. *J. Am. Chem. Soc.* **1980**, *102*, 4546–4548.

22. Garroway, A. N.; Ritchey, W. M.; Moniz, W. B. *Macromolecules* **1982**, *15*, 1051.

23. Wilson, R. J.; Meijer, G.; Bethune, D. S.; Johnson, R. D.; Chambliss, D. D.; de Vries, M. S.; Hunziker, H. E.; Wendt, H. R. *Nature (London)* **1990**, *348*, 621–622.

24. Wragg, J. L.; Chamberlain, J. E.; White, H. W.; Krätschmer, W.; Huffman, D. R. *Nature (London)* **1990**, *348*, 623–624.

25. Yannoni, C. S.; Bernier, P. P.; Bethune, D. S.; Meijer, G.; Salem, J. R. *J. Am. Chem. Soc.* **1991**, *113*, 3190–3192.

26. Engelsberg, M.; Yannoni, C. S. *J. Magn. Reson.* **1990,** *88,* 393.

27. Gutowsky, H. S.; Pake, G. E. *J. Chem. Phys.* **1950,** *18,* 162.

28. Carr, H. Y.; Purcell, E. M. *Phys. Rev.* **1954,** *94,* 630.

29. Meiboom, S.; Gill, D. *Rev. Sci. Instrum.* **1958,** *29,* 688.

30. Pake, G. E. *J. Chem. Phys.* **1948,** *16,* 327.

31. Feng, J.; Li, J.; Wang, Z.; Werner, M. C. *Int. J. Quant. Chem.* **1990,** *37,* 599.

32. Shibuya, T.-I.; Yoshitani, M. *Chem. Phys. Lett.* **1987,** *137,* 13–16.

Received September 12, 1991

Chapter 8

Mass Spectrometric, Thermal, and Separation Studies of Fullerenes

Donald M. Cox[1], Rexford D. Sherwood[1], Paul Tindall[1], Kathleen M. Creegan[1], William Anderson[2], and David J. Martella[3]

[1]Corporate Research Laboratories, Exxon Research and Engineering Co., Annandale, NJ 08801
[2]Department of Chemistry, Lehigh University, Bethlehem, PA 18018
[3]Exxon Chemicals Co., Linden, NJ 07036

In this chapter, we describe results from plasma desorption mass spectrometric studies of fullerenes, studies of the thermal and oxidative properties of the fullerenes, and the use of sublimation both for extraction of fullerenes from raw soot and for separation into purified fractions.

Following the initial disclosure of a technique to produce macroscopic quantities of C_{60} (*1*), the production, extraction, and separation of fullerenes is now being routinely performed by many research groups around the world, as indicated by the chapters in this book. There is a growing need for nondestructive mass spectrometric techniques with which parent fullerene molecular species can be identified, information about the thermal and oxidative properties of the fullerenes, and alternatives to liquid chromatographic separation (*2–4*). Considering these needs, we will discuss our preliminary results from plasma desorption mass spectrometric studies of evaporated deposits of purified and mixed fullerene samples, thermogravimetric studies of fullerenes and carbon soot with both N_2 and O_2 sweep gases, and application of sublimation–condensation not only to extract fullerenes from raw soot, but also as a means to separate fullerenes into highly purified components.

Plasma Desorption Mass Spectrometry

Figure 1 shows a schematic of plasma desorption mass spectrometry (PDMS) apparatus. Briefly, PDMS works as follows: A sample is placed on a thin film ($\sim$2 μm thick) of aluminized Mylar. The fissioning of a ^{252}Cf atom produces two high-energy, nearly equal mass products. One fission fragment produces a start pulse for data acquisition. The second fragment strikes the film, deposits

0097–6156/92/0481–0117$06.00/0

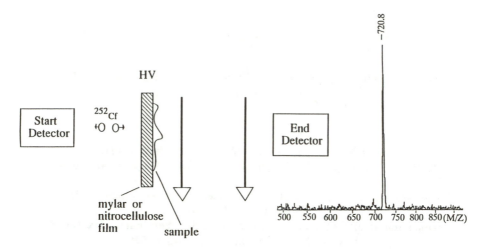

Figure 1. Schematic of plasma desorption mass spectrometer and the positive ion time-of-flight mass spectrum of purified C_{60} prepared by evaporative deposition of C_{60} from a dilute solution of toluene onto a thin nitrocellulose coating on the aluminized Mylar film. The accelerating voltage is +15 keV.

its kinetic energy, and produces ions. The exact details of the ion production are uncertain and the subject of much discussion (5). The sample ions are accelerated by an electric field set up between the substrate held at high voltage and a grounded screen about 2 mm away. The ions then travel in a field-free region for about 14 cm before being detected by a channel-plate multiplier. Simply changing the sign of the accelerating voltage allows either positive or negative ions to be detected.

This technique has advantages and disadvantages. One advantage is that fullerenes can be detected with this technique. Also shown in Figure 1 is a typical mass spectrum obtained from a "purified" sample of C_{60}. In this instance a low concentration of the fullerene molecule was evaporatively deposited onto the Mylar film from a weak solution of C_{60} in toluene. C_{60} is by far the dominant peak in the mass spectrum, a result indicating that the parent C_{60} molecules are being detected.

Figures 2a and 2b show spectra obtained when thicker films of "purified" C_{60} and raw fullerene extract (a fullerene mixture containing mostly C_{60} and C_{70}) are used, respectively. In both Figures 2a and 2b there is evidence for parent molecules (the strongest peaks), but also for "all" larger size fullerenes; that is, although not shown here, even atom carbon clusters containing 60 to more than 250 carbons are mass resolved. Our interpretation of these effects is that plasma desorption can be used to selectively detect parent fullerene molecules only when the fullerenes are highly dispersed on the film, that is, when the probability of the fission fragment interacting with more than one C_{60} molecule at a time is very low. For concentrated coverages, on the other hand, a significant fraction of the energy deposited by the fission fragment presumably is deposited into this "carbon film", and in analogy with laser vaporization of graphite (6), simply "synthesizes" the large family of fullerenes.

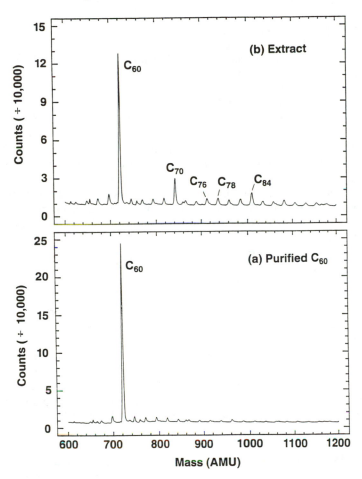

Figure 2. Positive ion time-of-flight mass spectra for more concentrated samples of (a) purified C_{60} and (b) fullerene extract. These samples were prepared by evaporative deposition onto aluminized Mylar films from concentrated solutions of fullerene in toluene. The accelerating voltage is +14 keV.

The PDMS of the fullerene extract shows higher intensities for C_{60}, C_{70}, C_{76}, C_{78}, and C_{84}, exactly those fullerenes that can be extracted by chromatography (2–4, 7). This finding suggests that such intensity oscillations are indicative of the presence of parent molecular species in the sample, because such oscillations are not present in the PDMS of a thick sample of purified C_{60} (or C_{70}). The most intense peak in this instance is C_{60} (or C_{70}), while the intensities of the other peaks are nearly equal except for a slow decrease with increasing cluster size.

Separation of Fullerenes

The disclosure (*1*) that fullerenes can be produced in macroscopic amounts by operating a carbon arc in an inert gas atmosphere has led to a demand for purified materials. The initial problems involved extraction of fullerenes from the raw soot and separation of the mixture of fullerenes into pure cuts of C_{60}, C_{70}, C_{84}, etc. A typical recipe for these steps involves combining the raw soot and a solvent such as benzene or toluene in a flask. The fullerenes are somewhat soluble in aromatic solvents, and thus after filtering out the insoluble material, a deep burgundy solution (the extract) containing only soluble material is obtained. This solution is found to contain a mixture consisting primarily of fullerenes, but may also contain other fullerenelike materials such as the fullerene mono- and dioxides and possibly some polyaromatic hydrocarbon species. Typically, under our "synthesis" conditions the fullerenes make up 8–15% of the carbon soot. In order to separate the fullerenes from each other, liquid chromatography has been employed (*2–4, 7*). In this manner, purified samples of C_{60}, C_{70}, C_{84}, etc., have been obtained.

In what follows we will describe appropriate sublimation conditions that not only allow fullerenes to be extracted from raw soot, and thus obviate the solvent-extraction step, but also yield highly purified deposits of C_{60} (and C_{70}), and thus obviate the chromatographic step. The method exploits the properties that C_{60} and C_{70} are thermally stable (above 800 °C in our tests) in vacuum or under inert gas, sublime without decomposing, and have different heats of desorption (*8*). The general behavior is that the larger fullerenes condense onto warmer surfaces than do the smaller ones. Thus, fullerene extraction and separation from the raw soot by controlling the heating and condensation conditions can be a viable alternative to solvent extraction and liquid chromatographic separation.

Thermal Properties of Fullerenes.   The thermal and oxidative properties of fullerene-containing soots as well as separated fullerenes were examined by thermogravimetric analysis (TGA), differential scanning calorimetry (DSC), and differential thermal analysis (DTA) using nitrogen, helium, or air. Figure 3 displays a typical TGA plot for a sample of toluene-Soxhlet-extracted fullerenes. With N_2, no appreciable weight loss is observed for temperatures less than ~700 °C. Most of the weight loss occurs between 750 and 850 °C. Little weight loss occurs above 850 °C, with approximately 13% of the initial fullerene sample still remaining even at 1100 °C. This residual material is insoluble in toluene and presumably represents conversion of fullerenes or fullerene oxides into some carbonaceous material, possibly larger fullerenes with more graphitelike properties, or amorphous carbon.

When synthetic air (20% O_2, 80% N_2) is used as the carrier gas, weight loss occurs at substantially lower temperatures; that is, almost all weight loss now occurs between 550 and 650 °C, and nearly all the material is oxidized, as evidenced by 98% weight loss. Similar studies of raw soot, that is, the carbon soot prior to fullerene extraction, show only about 10% weight loss with N_2 as the carrier gas, but with air the entire sample can be oxidized. The 10% loss

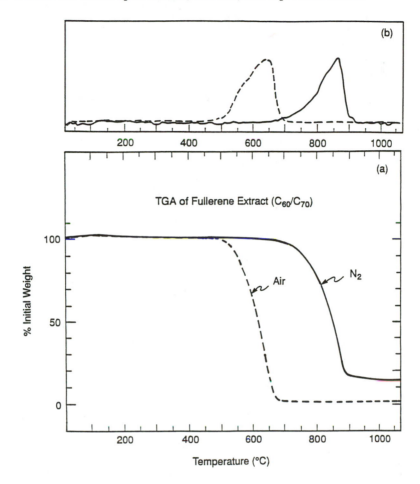

Figure 3. The TGA plots (% weight loss versus temperature) for the fullerene extract for N_2 (solid line) and synthetic air (dashed line, 20% O_2, 80% N_2) are compared in panel a. The heating rate was 20 °C/min. The upper panel (b) plots the derivatives of the curves in panel a.

with N_2 not only reflects the fraction of sublimable material (fullerenes), but is consistent with the weight fraction of fullerenes obtained by toluene extraction.

These results are interpreted as follows: With N_2 as the carrier gas, weight loss occurs by sublimation. With air as the carrier gas, weight loss occurs via oxidation (burning) before any significant weight is lost by sublimation. These results are confirmed by DTA and DSC on the extracted fullerenes. A typical DTA scan is shown in Figure 4. With N_2, no evidence of an endo- or exothermic reaction is observed. With air, however, the reaction is exothermic. The shape of the curve suggests that the sample contains two major components, presumably C_{60} and C_{70}. In addition, we suspect that the shoulder toward the lower temperature reflects the oxidation of C_{70}, and the

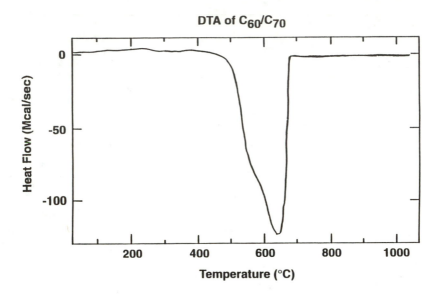

Figure 4. DTA of the fullerene extract in air. The heat flow (millicalories per second) is plotted such that exothermic processes give negative values.

large peak corresponds to oxidation of C_{60}, because the ratio of the signal in the shoulder to that of the peak is consistent with the C_{70} to C_{60} ratio of about 0.2 in this sample. Furthermore, a similar effect is observed in the derivative of the TGA with oxygen (*see* Figure 3b). If this assignment is correct, then C_{70} oxidizes more easily than C_{60}. Early mass spectrometric characterizations of fullerene extracts in many instances were carried out by thermal desorption from heated probes where probe temperatures were held between 300 and 500 °C (*1, 2–4, 9*), temperatures significantly below that for which significant weight loss is observed. In these instances the mass spectrometric sensitivity is sufficiently high that even at the lower vapor pressures sufficient material is sublimed for detection.

Separation by Sublimation. In order to test the feasibility of separation using sublimation, an initial experiment is carried out by placing a known amount of fullerene extract (toluene-extracted as usual) into the bottom of a ¼-inch-diameter by 12-inch-long quartz tube. The tube is evacuated and sealed off. It is then placed in an oven such that a 4-inch extension out of the end of the oven remains near room temperature. The tube containing the soot is then heated to 700 °C for 60 h. A thick deposit is formed on the inner surface of the tube over about a 2-inch-long region located where the tube exits the oven, that is, the region where the temperature drops from near 700 °C to somewhere near room temperature. Characterization of this deposit reveals that the material nearest the cool end is highly enriched in C_{60}, containing greater than 95% C_{60}. The percent C_{60} in the extract prior to sublimation is 83%. Further evidence of selective sublimation is obtained by examination of the extract that did not sublime. This material is found to be significantly depleted in C_{60}, containing only 70% C_{60}.

To better define the thermal conditions and to demonstrate that fullerenes can be extracted and separated directly from soot, a three-zone oven is used. Figure 5 schematically shows the details of this setup. Each stage of the oven is nominally 6 inches long and has a thermocouple located near the center of each stage. Raw carbon soot (200 mg) is placed in the bottom of a 24-inch-long by 0.5-inch-diameter quartz tube, which is then positioned in the oven such that the lower 5 inches containing the soot is in the 800-°C zone. The three zones of the oven are then heated to 800, 600, and 400 °C. The top 8 inches of the tube outside the oven remains near room temperature.

Our observations are as follows: No appreciable material is deposited on the room-temperature surfaces outside the oven or in the 600-°C region. A coating builds up along the tube surface in the 400-°C region. The morphology of the coating varies along the tube length and is found to reflect both compositional (C_{60}/C_{70} ratio changes) and concentration (film thickness) changes. The differing compositions and concentrations of sublimed material appear to be controlled mostly by the temperature profile along the length of the quartz tube. The deposited material is then removed from different areas along the length of the tube, weighed, and characterized by UV absorption. From these measurements the percent C_{60} and the yield of sublimable material is determined (*see* Table I). The two spectra shown at the bottom of Figure 5 dramatically illustrate the differing concentrations of C_{60} and C_{70} in the cooler and hotter regions along the tube.

Table I. Quantity (%) of C_{60} and C_{70} in Sublimed Films as a Function of Position along Length of Quartz Tube

Position	C_{60}	C_{70}[a]
A_{1-2}	>98	<2
B_{1-3}	~80	~20
B_4	~24	~76
C	~12	~88

NOTE: *See* Figure 5 for explanation of positions.

[a]The larger fullerenes are present in our soot at levels less than 1–2%; therefore, we assume only C_{60} and C_{70} contribute to the UV absorption at 330 and 380 nm. Thus, %C_{70} = 100 − %C_{60}.

Last, the total yield of fullerenes (sublimed material that is also toluene soluble) is about 10%, a value nearly identical with that obtained by solvent (toluene) extraction on another sample of soot from the same run. Examination of the results from this run are summarized in Table I and show that highly purified C_{60} and C_{70} can be obtained by sublimation, and sublimation under appropriate conditions is at least as effective as solvent extraction in extracting the smaller fullerenes from soot.

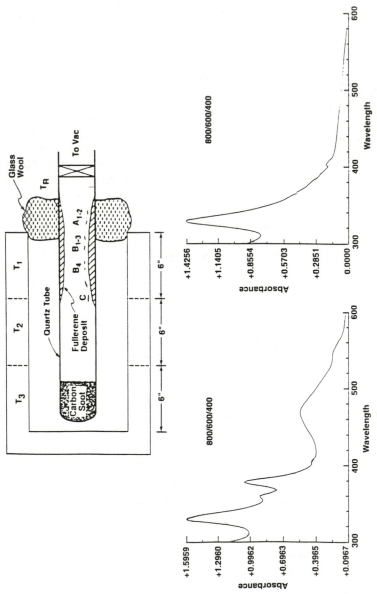

Figure 5. Schematic of three-stage oven and UV spectra for fullerene material deposited at two different regions (temperatures) along the length of the 0.5-inch-diameter quartz tube. The UV spectrum for material collected from region A_{1-2} is shown on the right, and that from region B_4 on the left. The ratio of the peaks near 330 and 380 nm is used to calculate the C_{60} to C_{70} concentrations (see Table I).

Summary

We have described the characteristics of plasma desorption mass spectrometric detection of fullerene samples. In addition we have shown that the thermal and oxidative properties of the fullerenes can be exploited for extraction and preparation of purified C_{60} and C_{70} by controlled sublimation.

References

1. Krätschmer, W.; Lamb, L. D.; Fostiropoulos, K.; Huffman, D. R. *Nature (London)* **1990**, *347*, 354–358.

2. Taylor, R.; Hare, J. P.; Abdul-Sada, A. K.; Kroto, H. W. *J. Chem. Soc. Chem. Comm.* **1990**, 1423–1425.

3. Ajie, H.; Alvarez, M. M.; Anz, S. J.; Beck, R. D.; Diederich, F.; Fostiropoulos, K.; Huffman, D. R.; Krätschmer, W.; Rubin, Y.; Schriver, K. E.; Sensharma, D.; Whetten, R. L. *J. Phys. Chem.* **1990**, *94*, 8630–8633.

4. Cox, D. M.; Behal, S.; Disko, M.; Gorun, S. M.; Greaney, M.; Hsu, C. S.; Kollin, E. B.; Miller, J.; Robbins, J.; Robbins, W.; Sherwood, R. D.; Tindall, P. *J. Am. Chem. Soc.* **1990**, *113*, 2940–2944.

5. For example see, Fenyo, D; Sundqvist, B. U. R.; Karlsson, B. R.; Johnson, R. E. *Phys. Rev.* **1990**, *B42–45*, 1895.

6. Rohlfing, E. A.; Cox, D. M.; Kaldor, A. *J. Chem. Phys.* **1984**, *81*, 3322–3330 .

7. Diederich, F.; Ettl, R.; Rubin, Y.; Whetten, R. L.; Beck, R.; Alvarez, M.; Anz, S.; Sensharma, D.; Wudl, F.; Khemani, K. C.; Koch, A. *Science (Washington, D.C.)* **1991**, *252*, 548–550.

8. Pan, C.; Sampson, M. P.; Chai, Y.; Hauge, R. H.; Margrave, J. L. *J. Phys. Chem.* **1991**, *95*, 2944–2946.

9. Haufler, R. E.; Conceicao, J.; Chibante, L. P. F.; Chai, Y., Byrne, N. E.; Flanagan, S.; Haley, M. M.; O'Brien, S. C.; Pan, C.; Xiao, Z.; Billups, W. E.; Ciufolini, M. A.; Hauge, R. H.; Margrave, J. L.; Wilson, L. J.; Curl, R. F.; Smalley, R. E. *J. Phys. Chem.* **1990**, *94*, 8634–8636.

Received September 11, 1991

Chapter 9

Production, Mass Spectrometry, and Thermal Properties of Fullerenes

Ripudaman Malhotra, Donald C. Lorents, Young K. Bae,
Christopher H. Becker, Doris S. Tse, Leonard E. Jusinski,
and Eric D. Wachsman

Chemistry and Molecular Physics Laboratories, SRI International,
Menlo Park, CA 94025

In this chapter, we report on the generation of gram quantities of solid molecular carbon fullerene materials (fullerites) by a modified version of the recently discovered technique of Krätschmer et al.; we used an alternating-current arc instead of resistive heating. The presence of fullerenes in the soot was confirmed by surface analysis using laser ionization (SALI) and field-ionization mass spectrometry (FIMS), both of which showed the parent ions at 720 and 840 Da. The absence of these peaks in the SALI and FIMS spectra of commonly encountered pyrolysis and combustion soots leads to the conclusion that fullerenes are not present in those samples. C_{60} desorbed at lower temperatures than C_{70} from the raw soot obtained by vaporizing carbon; however, the two fullerenes coevolved during thermal desorption of bulk fullerene extract. Furthermore, at temperatures above 700 °C, the fullerenes transformed to another form of carbon that is insoluble in benzene. The kinetics of oxidation of C_{60} and C_{70} were found to be very similar by temperature-programmed reaction studies and somewhat faster than for graphite.

Completely new classes of materials can be formed from size-selected metal and semiconductor clusters, so-called "nanophase" structures. Among clusters with natural size selection, C_{60} is now recognized as a unique molecule with its truncated icosahedron (soccer-ball) shape determined by the pentagonal and hexagonal framework of C–C bonds. It has been shown to be chemically, thermodynamically, and photophysically stable in the gas phase (*1, 2*). Even-numbered carbon fullerene structures larger than C_{32} have been observed, but except for C_{60}, only those larger than C_{70} appear to be stable. Bulk materials composed of these carbon fullerene molecules can be expected to have interesting and unique solid-state properties and applications (*1, 2*). Although there have been extensive gas-phase studies of these fullerenes (*1–3*), studies of their material properties have been limited by the difficulty of producing bulk quantities of fullerenes.

0097–6156/92/0481–0127$06.00/0

Recently, Krätschmer et al. (*4*) discovered a simple production and extraction method that permits bulk quantities of fullerenes to be produced in the solid form, called fullerite. Fullerite is a completely new form of carbon that, unlike its counterparts diamond and graphite, is soluble in common solvents and volatile at low temperatures. Fullerites composed of pure C_{60} or pure C_{70} can now be produced by liquid chromatographic separation techniques (*5, 6*). This form of carbon provides a basis for the development of new classes of technologically important materials, such as new semiconductor materials, new organic materials produced by chemical functionalization of fullerenes, and new high-temperature coatings and lubricants. In this volume, Smalley presents a more general discussion on this subject. In this chapter we describe measurements of some of the thermal properties of fullerite materials. We had presented some of these results previously (*7*).

Production

We have generated gram quantities of fullerenes, such as C_{60} and C_{70}, by extracting them from carbon smoke deposits with benzene or toluene following the procedure discovered by Krätschmer et al. (*4*). The smoke deposits were initially produced by thermal evaporation of graphite rods in He at ~150 torr ($\sim 2 \times 10^4$ Pa). In the course of these experiments we found that operation in an alternating-current (AC) arc discharge mode is both simpler and more efficient than resistive heating. A typical arc operation at ~200 A and ~20 V permits one to vaporize 6-inch-long, ¼-inch-diameter carbon rods in an ~2-h run. A 4–5-g carbon soot deposit on a water-cooled copper collector is produced from a single rod and yields ~10 wt% fullerenes.

Our arc discharge method appears to be identical to that described by the Rice University group (*8*). Typically, we extract the soot by sonicating it with about 200 mL of toluene and filtering the extract. Often the soot produces a sticky solid with toluene, and in those instances filtration is particularly problematic. We found that filtration with a 12.5-cm glass-fiber filter disc followed by a second stage of filtration through a 0.45-μm membrane filter minimized clogging problems. The initial residue obtained after evaporation of the solvent is black. Besides the characteristic absorption bands at 1183 and 1429 cm^{-1}, Fourier transform infrared (FTIR) spectroscopy shows the presence of peaks in the 3000-cm^{-1} range, which indicates hydrocarbon contaminants. Simple washing with ether or chloroform removes these impurities, as is verified by FTIR spectroscopy, and the cleaned solid has a brownish tinge.

Mass Spectrometry of Fullerenes

We analyzed the raw soot and the extracts by two different mass spectrometric methods: (1) surface analysis by laser ionization (SALI) and (2) field-ionization mass spectrometry (FIMS). SALI analysis was performed by heating

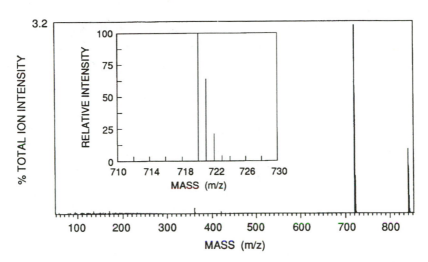

Figure 1. Field-ionization mass spectrum of soot obtained by vaporizing graphite shows peaks due to C_{60} and C_{70}. Inset: Expanded region around mass 720 showing the isotopic satellites.

the sample on a stainless steel stage and photoionizing the evaporating fullerenes and other molecules with 118-nm (10.5-eV) photons. The ions were then monitored with time-of-flight (TOF) mass spectrometry. The laser photons were generated by first tripling the 1.06-mm photons in KD*P, then tripling again in Xe phase-matched with Ar (9). The estimated ionization potentials of the small fullerenes, such as C_{60}, are 7–8 eV (10); thus, the photoionization process at 10.5 eV should be dominated by the single-photon process. Unlike multiphoton or electron ionization, single-photon ionization produces a clean, nearly fragment-free mass spectrum (9). Assuming that the photoionization efficiencies of the different fullerenes are equal, the resulting ion mass spectra represent the true intensity distributions of neutral fullerenes. The spectra clearly showed strong peaks at 720 and 840 Da, corresponding to C_{60} and C_{70}, respectively. This technique was also used to study the thermal desorption of fullerenes (discussed later).

For FIMS analysis, the sample was taken in a melting point capillary and introduced into the system via a heatable direct insertion probe (11). Ionization of evaporating molecules was effected with a foil-type field ionizer previously activated with carbon dendrites. The ions were analyzed with a 60° magnetic sector mass analyzer. Fullerenes were detected around 400 °C, although intense signals were observed only above 500 °C.

Figure 1 shows the FI mass spectrum of a typical extract. No fragment peaks were formed by this method. Weak signals (~0.3%) were detected at m/z 360 and 420 Da. They result from doubly charged species and are not due to clusters containing 30 and 35 carbons, for they coevolve with the peaks at 720 and 840 Da. The inset in Figure 1 displays the region around 720 Da and shows the presence of isotopic satellites due to ^{13}C in the expected ratios.

The existence of fullerenes in combustion and pyrolysis soot has been a matter of debate (2, 12–14). We had previously analyzed many pyrolysis and combustion soots, but had not seen any peak corresponding to C_{60} or C_{70}. However, because we could not be sure that our technique would detect them, the mere absence of FIMS signals was not a sufficient criterion to conclude the absence of fullerenes in these samples. However, the successful analysis of fullerenes by FIMS now allows us to categorically state that fullerenes such as C_{60} and C_{70} are not present in commonly encountered soots from pyrolysis and combustion of hydrocarbons (15). In a similar vein, we also found no evidence for fullerenes in the Allende and Murchison meteorites (16).

Recently Taylor et al. (17) showed that fullerenes react with oxygen in the presence of UV light, and they suggested that this reaction explains their absence in the combustion soots, the clear implication being that the combustion soot contained fullerenes that were subsequently transformed. Although we have not systematically studied the decay of fullerenes in carbonaceous particles, we analyzed some of the samples of pyrolysis and combustion soots within days of their preparation, and some soots from evaporating graphite in an arc discharge that were stored under ambient conditions for weeks prior to mass spectrometric analysis. Thus it seems unlikely that fullerenes were originally present in the samples of combustion soot. This result does not mean that fullerene cannot be prepared by combustion processes. Calculations by McKinnon (18) and recent experimental results of Howard et al. (19) indicate that under specific conditions fullerenes can be formed during combustion in reasonable yields.

Thermal Desorption of Fullerenes

We examined the thermal desorption properties of fullerenes from the raw carbon smoke deposit and from bulk fullerenes with the SALI technique. Evaporation of bulk fullerenes and pure C_{60} samples were studied by thermogravimetry.

Desorption from Raw Soot. The temperature-dependent SALI spectra of the volatile compounds from the raw soot obtained from the arc discharge are shown in Figure 2. For these spectra, the sample was heated at a constant rate of ~10 °C/min. Figure 2a exhibits some low-mass components (<300 amu) at temperatures below 500 °C, which quickly disappear as the temperature increases above 500 °C. We believe that these low-mass components are hydrocarbon contaminant molecules generated in the discharge. C_{60} and C_{70} appear first at about 400 °C and become predominant at temperatures between 500 and 700 °C as shown in Figures 2a–2c. As Figures 2a–2d clearly indicate, no fullerenes smaller than C_{60} or between C_{60} and C_{70} appear in these spectra. However, C_{78} begins to appear at 700 and 800 °C, as shown in Figure 2d; the higher mass fullerenes C_{74}, C_{76}, C_{78}, C_{82}, and C_{84} also appear in the 800-°C spectrum. Among them, C_{76}, C_{78}, and C_{84} are predominant, a feature indicating their formation and stability.

A general trend occurs in the temperature dependence of thermal desorption as a function of fullerene size. The lowest member, C_{60}, is the first to desorb, C_{70} is next, and the higher masses appear only at the highest temperature. The intensity of C_{60} (Figure 2b) peaks at about 600 °C, and that of C_{70} (Figure 2d) is still increasing at 800 °C. This result is in concert with the findings of Cox et al. *(21)*; however, as shown later, from bulk fullerite the desorption of C_{60} and C_{70} occurs over the same temperature range.

Desorption from Bulk Fullerene Extract. Thermal desorption properties of the solvent extract (composed mainly of C_{60} and C_{70}) using SALI analysis are displayed in Figure 3a, which shows the signal intensity of C_{60} and C_{70} as a function of temperature from two different sample preparations. Curves C and D were derived from a sample prepared by evaporating the equivalent of one or two monolayers of fullerenes on the Ta heating strip by air-drying 3 mL of a dilute fullerene solution in toluene. Curves A and B were prepared in the same manner but with about 5–10 monolayers deposited on the strip. The exact microscopic surface area of the Ta was unknown. The samples were placed into the vacuum chamber through a vacuum interlock. The heating rate was 30 °C/min. The low-temperature peaks or shoulders of the curves at ~300 °C are probably due to evaporation of monolayer fullerenes from the TaO surface and reflect a rather weak binding to this oxide. The second set of peaks at just below 400 °C are believed to arise from multilayers of fullerenes. The rapid falloffs above the second peak (just above 400 °C) in all curves result from sample exhaustion.

Figure 3b shows the thermal evolution from a flake of unpurified molecular carbon extract (dried solvent extract) material (about 70 mg) placed in a fine Cu mesh and tied by a fine Ta wire to the Ta strip heater. The heating rate was 15 °C/min. In this case the predominant evaporation occurs at much higher temperatures (500–700 °C). However, a weak shoulder that appears at about 400 °C corresponds to the multilayer peaks of Figure 3a. A surprising feature of these curves is the observed limiting of the vaporization rate at temperatures above about 550 °C. This decrease in desorption rate may indicate a transformation that increases the heat of vaporization. These observations indicate that the desorption processes from films and crystallites are quite different. Different desorption sites, different orders of desorption, and/or phase changes may be involved.

All of the curves in Figure 3 clearly show that the C_{60} and C_{70} have nearly identical thermal evolution. Furthermore, the evolution curves of C_{78} and C_{84} are nearly identical to the C_{60} curves. This result is in marked contrast to the results for thermal evolution from the raw carbon deposit where the higher mass fullerenes vaporize at higher temperatures than the smaller ones. The thermal evolution from the raw deposit appears to be associated mainly with desorption of the fullerenes from the larger carbon structures in the deposit. However the thermal evolution from thin films and mixed fullerene crystals is apparently dominated by fullerene–fullerene interactions that are similar for C_{60} and C_{70}.

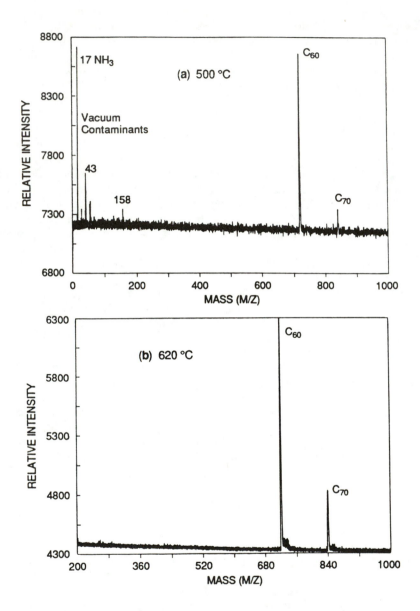

Figure 2. SALI spectra of fullerenes evaporated from the raw soot at different temperatures. The fullerenes were photoionized with 118-nm photons. (Reproduced with permission from reference 20. Copyright 1991 Materials Research Society.)

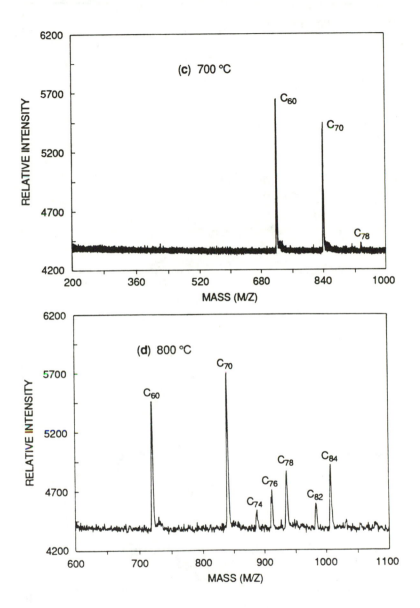

Figure 2. Continued.

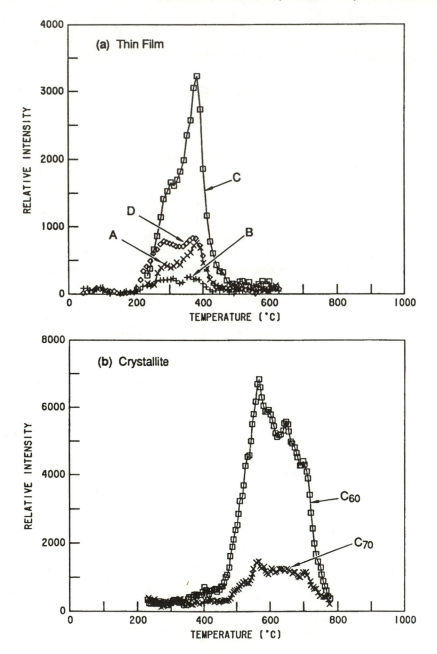

Figure 3. Thermal evolution profiles of C_{60} and C_{70} fullerenes from thin films and crystallites of mixed fullerite. (Reproduced with permission from reference 20. Copyright 1991 Materials Research Society.)

Thermogravimetric Analysis (TGA)

Additional information about the thermal vaporization properties of these extracts, both purified and unpurified, was obtained from temperature-programmed weight-loss (thermogravimetric) measurements. Initial measurements were made on a mixture of C_{60} and C_{70} in flowing Ar. Two noteworthy observations from these measurements are (1) the temperature of the peak vaporization rate depends on heating rate, and (2) the vaporization curve appeared to be composed of two components. The possibility that these features were due to the admixture of C_{70} or other fullerenes in the sample led us to repeat the measurements on pure C_{60} samples, which also showed the same behavior.

Figure 4a shows the result obtained by ramping (slowly increasing) the temperature of the pure C_{60} sample at a rate of 25 °C/min. The sample begins

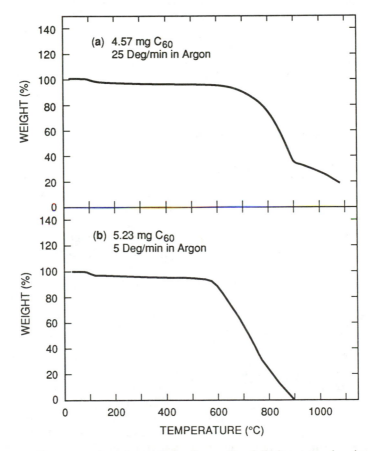

Figure 4. Thermogravimetric analysis of pure C_{60} fullerite at two heating rates: (a) 25 °C/min and (b) 5 °C/min. A significant slowing of the weight loss occurs near 900 °C in plot a and near 750 °C in plot b. (Reproduced with permission from reference 20. Copyright 1991 Materials Research Society.)

to lose weight significantly above 600 °C because of evaporation of C_{60}. However, the rate of weight loss decreases abruptly around 900 °C. At a slower heating rate (5 °C/min) the weight loss due to evaporation was observed above 550 °C, and a change in the rate, albeit less pronounced, occurred near 750 °C. The observed dependence on heating rate is expected for any vaporization process, and in principle one can extract the heat of vaporization from these data. Analysis of the front end of the individual TGA curves using either zero- or first-order kinetics gives heat of vaporizations around 25 kcal/mol, a value consistent with that reported by Haufler et al. (8). On the other hand, if we use the difference in the peak temperatures for vaporization obtained at the two heating rates, the calculated heat of vaporization is on the order of 55 kcal/mol.

Evidently, the sample is not undergoing simply a vaporization process, but also some other changes are occurring. The sudden change in the vaporization rate observed at high temperatures is further evidence of a thermally activated transformation of the fullerenes into some new form of carbon. Although this new species evaporates at relatively low temperature compared to graphite (~4000 °C), it is insoluble in benzene. We conjecture that this new carbon may be composed of larger fullerene structures that have thermally bonded together, possibly through Diels–Alder like adducts. In a recent study, Ismail (22) also reported formation of an insoluble residue during TGA. Further investigation of this process is under way.

Temperature-Programmed Oxidation

The susceptibility of fullerenes toward oxidation was investigated by temperature-programmed reaction (TPR) in flowing mixtures of O_2 and He. The product gases CO and CO_2 and remaining oxygen were monitored with a quadrupole mass spectrometer. Figure 5a shows the evolution of CO and CO_2 from pure C_{60} (~5 mg) heated at 30 °C/min in a stream of 10.3% O_2 in He flowing at 30 cm^3/min. The TPR curve shows only a single peak at 550 °C, with the onset of oxidation taking place at about 400 °C. Similar results were obtained for the TPR of C_{70} under the same experimental conditions. Analysis of the front end of the curves gives apparent activation energies of 39 and 24 kcal/mol, respectively, for C_{60} and C_{70}. This difference in apparent activation energies is significant and is under further investigation.

Figure 5b shows the TPR curve for spheron, a graphitized carbon, under conditions identical to those used for C_{60} and C_{70}. From the onset of oxidation at 450 °C for spheron, we can conclude that fullerenes are less resistant to oxidation than graphite. Ismail (22) and Milliken et al. (23) also arrived at the same conclusion from their TPR studies. Another feature worth noting is that the CO_2 to CO ratio produced in the oxidation of fullerenes is about 1.2, whereas that produced from spheron is about 5.6. The large difference in the CO_2/CO ratio cannot be simply due to the gas-phase reactions of carbon oxides and oxygen, because this difference persists even in the temperature range common to both systems. This ratio probably reflects the differing nature of the

(a)

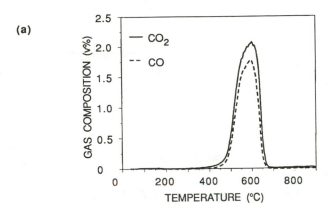

(b)

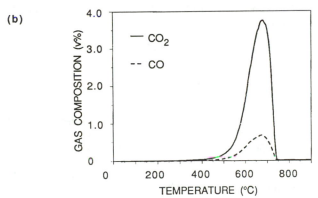

Figure 5. Temperature-programmed oxidation of (a) pure C_{60} and (b) spheron, a graphitized carbon. Sample size, ~5 mg; 10.3% O_2 in He; heating rate, 30 °C/min.

intermediates involved the gasification. However, at present we are unable to draw any mechanistic conclusion from this observation.

Summary

In summary, we produced and characterized bulk materials formed from fullerene carbon clusters. We investigated thermal desorption properties of fullerenes from the raw carbon smoke deposit, from thin fullerite films, and from solvent-grown crystallites. The desorption characteristics of fullerenes from the raw smoke are very different from the fullerite. The vaporization and desorption processes are complex and reflect possible thermal transformations to larger fullerene structures, and fullerenes are less resistant to oxidation than graphite.

Acknowledgments

We thank A. McEwen, D. Keegan, and M. Dyer for their technical contributions. This research was supported by the Independent Research and Development Program of SRI International.

References

1. Curl, R. F.; Smalley, R. E. *Science (Washington, D.C.)* **1988**, *242*, 1017, and references therein.

2. Kroto, H. *Science (Washington, D.C.)* **1988**, *242*, 1139, and references therein.

3. Cox, D. M.; Reichmann, K. C.; Kaldor, A. *J. Chem. Phys.* **1988**, *88*, 1588, and references therein.

4. Krätschmer, W.; Lamb, L. D.; Fostiropoulos, K.; Huffman, D. R. *Nature (London)* **1990**, *347*, 354.

5. Taylor, R.; Hare, J. P.; Abdul-Sada, A. K.; Kroto, H. W. *J. Chem. Soc. Chem Commun.* **1990**, 1423.

6. Ajie, H.; Alvarez, M. M.; Anz, S. J.; Beck, R. D; Diederich, D.; Fostiropoulos, K.; Huffman, D. R.; Krätschmer, W.; Rubin, Y.; Schriver, K. E.; Sensharma, D.; Whetten, R. *J. Phys. Chem.* **1990**, *94*, 8630.

7. Bae, Y. K.; Lorents, D. C.; Malhotra, R.; Becker, C. H.; Tse, D. S.; Jusinski, L. *Mater. Res. Soc. Symp. Proc.* **1990**, *205*, 733.

8. Haufler, R. E.; Conceicao, J.; Chibante, L. P. F.; Chai, Y.; Byrne, N. E.; Flanagan, S.; Haley, M. M.; O'Brien, S. C.; Pan, C.; Xiao, Z.; Billups, W. E.; Ciufolini, M. A.; Hauge, R. H.; Margrave, J. L.; Wilson, L. J.; Curl, R. F.; Smalley, R. E. *J. Phys. Chem.* **1990**, *94*, 8634.

9. Pallix, J. B.; Schüle, U.; Becker, C. H.; Huestis, D. L. *Anal. Chem.* **1990**, *61*, 805.

10. Larsson, S.; Volosov, A.; Rosen, A. *Chem. Phys. Lett.* **1987**, *137*, 501.

11. St. John, G. A.; Buttrill, S. E., Jr.; Anbar, M. "Field Ionization and Field Desorption Mass Spectrometry Applied to Coal Research," In *Organic Chemistry of Coal*; Larsen, J. W., Ed.; ACS Symposium Series 71, American Chemical Society: Washington, DC, 1978.

12. Zhang, Q. L.; O'Brien, S. C.; Heath, J. R.; Liu, Y.; Curl, R. F.; Kroto, H.; Smalley, R. E. *J. Phys. Chem.* **1986**, *90*, 525.

13. Frenklach, M.; Ebert, L. B. *J. Phys. Chem.* **1988**, *92*, 563.

14. Ebert, L. B. *Science (Washington, D.C.)* **1990**, *247*, 1468.

15. Malhotra, R.; Ross, D. S. *J. Phys. Chem.* **1991**, *95*, 4599.

16. Tingle, T. N.; Becker, C. H.; Malhotra, R. *Meteoritics* **1991**, *26,* 117.

17. Taylor, R.; Parsons, J. P.; Avent, A. G.; Rannard S. P.; Dennis, T. J.; Hare, J. P.; Kroto, H.; Walton, D. R. M. *Nature (London)* **1991**, *351,* 277.

18. McKinnon, J. T. *J. Phys. Chem.,* accepted, 1991.

19. Howard, J. B.; McKinnon, J. T.; Makarovsky, Y.; Lafleur, A.; Johnson, M. E. *Nature (London)* **1991**, *352,* 139.

20. Bae, Y. K.; Lorents, D. C.; Malhotra, R. *Mater. Res. Soc. Symp. Proc.* **1991**, *206,* 738–740.

21. Cox, D. M.; Behal, S.; Disko, M.; Gorun, S.; Greaney, M.; Hsu, C. S.; Kollin, E.; Miliar, J.; Robbins, J.; Robbins, W.; Sherwood, R.; Tindall, P. *J. Am. Chem. Soc.* **1991**, *113,* 2940.

22. Ismail, I. M. K. *Am. Chem. Soc. Div. Fuel Chem. Prepr.* **1991**, *36*(3), 1026.

23. Milliken, J.; Keller, T. M.; Baronavski, A. P.; McElvany, S. W.; Callahan, J. H.; Nelson, H. H. *Chem Mater.* **1991**, *3,* 386.

Received August 30, 1991

Chapter 10

Doping the Fullerenes

R. E. Smalley

Rice Quantum Institute and Departments of Chemistry and Physics, Rice University, Houston, TX 77251

Fullerenes are a new class of carbon molecules, the first truly molecular form of pure carbon yet isolated. Consisting of hollow cages composed of three connected networks of carbon atoms arranged to form 12 pentagons and a varying number of hexagons, these spheroidal molecules may be most useful when they are mixed with small numbers of other atoms. These dopant atoms may be located (1) outside the cage, producing fulleride salts; (2) inside the cage, producing a sort of superatom; or (3) as part of the cage itself, replacing one or more of the carbon atoms in the cage network. Examples of all three types of doping have already been demonstrated.

For most observers, on first learning of the proposed soccer-ball structure for C_{60}, there often is an almost irrepressible urge to ask "Can you put something inside?" Here is a hollow molecule with a sort of holier-than-thou sphericity and symmetrical perfection. Inside is a perfect void. Perhaps it is natural to the human spirit, perhaps even healthy, to immediately think of messing up this sphere: changing just one of the carbon atoms for some other, and/or filling the void. The phrase, "Let's dope the bucky ball!" seems rather enticing.

There may also be considerable virtue in learning to do just that. In semiconductor technology and solid-state physics, doping an otherwise pure and pristine semiconductor lattice is what gives the material its vital electrical properties. For example, doping is what makes silicon work. Using such techniques as ion implantation, modern microelectronics companies add small, but controlled amounts of boron or phosphorus. Without these critical impurities, elegant perfect crystals of silicon would be useless insulators at room temperature.

The English–American word "dope" stems from the Dutch word "doop", which means sauce. In turn this doop is derived from earlier Dutch, German, and French usage of similar-sounding words for dipping in various fluids, and in particular, the rite of baptism. In Christianity this baptism is an act of cleansing in preparation for receipt of the life force of the Holy Spirit. For semiconductors and fullerenes, it's not immediately clear what this kind of doping would mean, but it certainly sounds like something worthwhile.

So how do we, following this etymology, go about baptizing C_{60}? Calculations (1, 2) and detailed experiments (3, 4) have revealed that crystalline films

0097–6156/92/0481–0141$06.00/0

of pure C_{60} are fcc (face-centered cubic) lattices that, in bulk form, are direct band-gap semiconductors, with a gap of roughly 1.7 eV. At room temperature this gap energy is many times the average thermal energy, kT, and hardly any electrons have enough energy to jump from ball to ball. At room temperature C_{60} is, therefore, effectively an insulator, just like pure silicon. But like silicon, it should be possible to dope the pure C_{60} lattice with balls that rather more readily give up, or take up, electrons. Once these extra electrons (or holes) travel far enough from their original modified-bucky home, they are screened from the charge they left behind, and they will serve as free carriers of electricity. In other words, it should be possible to produce n-type (electron carrier) or p-type (hole-carrier) doped C_{60} films, if we could only figure out a means of producing fullerenes with varying electronegativities.

There can only be three distinct ways of doing this. We can modify a fullerene like C_{60} (1) on the outside, (2) on the inside, or (3) on the side itself. As long as we deal with three-dimensional Euclidean space, these are the only possibilities. Examples of the feasibility of all three methods have already been published (5–9), although only the first has so far been used to produce interesting new materials in the laboratory. By mixing alkali metals such as rubidium and cesium into the fcc lattice, the superconducting alkali fulleride crystals are made (10), and a vast literature is growing about such externally doped fullerene materials.

This short chapter deals, instead, with the other two distinct ways of doping. Unlike the first, they constitute new molecular forms of fullerenes. In some cases they will be electronically closed-shell species, in many ways just like the pure, empty fullerenes themselves. But in most cases, these doped fullerenes, or "dopy balls" will be rather reactive: They will either be open-shell species like $C_{59}B$, or closed-shell singlets with close-lying, open-shell triplet diradicals. In many cases, these open-shell or nearly-open-shell species will be the most interesting and the most technologically worthwhile. But their synthesis and manipulation will be somewhat more of a challenge.

Nomenclature and Symbolism

These internally doped fullerenes are, in fact, new molecules, and therefore it will soon be necessary to evolve a systematic way of referring to them both by name and with symbols in a molecular formula. To chemists who have to confront the problem of systematizing myriads of possible molecules, the naming and representing of compounds is an important unifying discipline, which is not to be taken lightly. Ultimately, the IUPAC committee will have to consider this problem in detail, but for now we need to at least make a trial start. My colleagues and I have considered this problem rather extensively, and at least within our laboratory, have found the following symbolism to be expressive and nicely concise.

We have here a new class of molecules that are roughly spheroidal. They have enough room inside to house at least one atom of any element in the periodic table, and a number of substitutions are feasible for the carbon atoms in the cage itself. In a useful sense, these complex fullerenes are superatoms, and it would be nice to have the symbolism represent this fact. We therefore use a set of parentheses to group the relevant atom symbols together, and we use the "at" symbol, @, to denote that these atoms will make a fullerene. Atoms that are located in the central cavity inside the fullerene cage are grouped to the left of this @ symbol, whereas atoms that are part of the cage are listed to the right. For example, the archetypical fullerene, C_{60}, is then explicitly written as $(@C_{60})$. This convention seems needlessly complex until we consider how to represent a 60-cage fullerene with a potassium on the inside, several potassium atoms on the outside, and one boron substituting for carbon in the cage. This molecule has actually been synthesized in small amounts as a positive ion levitated in an magnetic field (*11*). It takes a lot of words to describe it, but with the @ symbolism, it is easily expressed as $K_2(K@C_{59}B)$.

Most likely, chemists in general will respond to this suggestion with some healthy conservatism—why bring in an entirely new symbol to a way of writing chemical formulas that has been perfectly satisfactory for millions of other compounds over the past many decades? The answer is that we don't have to. We do need some sort of single character delimiter to distinguish those atoms that are inside the cage, and those that make up the cage itself. But we could use a comma: $K_2(K,C_{59}B)$, or a period: $K_2(K.C_{59}B)$, or $K_2(K\text{-}C_{59}B)$, or $K_2(K>C_{59}B)$, or subscripts, or superscripts, etc. But none of these serves to suggest the nature of the object being specified nearly so well as $K_2(K@C_{59}B)$. The @ symbol is an almost perfect picture of a central atom surrounded by a spheroidal cage. It is a standard character available on all modern keyboards, readily printed and transmitted electronically. We need something that says the atoms to the left are to be found at the center of the fullerene cage made from the atoms following on the right. Why not use the symbol for "at"?

The more complex question of systematic names for these complex fullerene molecules has been left alone for now. In what follows a few new names such as "borofullerene" are tried out, but no attempt at a systematic nomenclature has yet been attempted.

Doping the Cage

So how do we dope the cage? Thus far only one technique has been proven effective, and not in high yield. But it is a start. It is a generalization of the original method (*12*) for generating C_{60}. A short, high-energy laser pulse is used to vaporize a graphite target, producing a carbon plasma that rapidly condenses in a pulse of helium to produce small clusters. When this cluster distribution is examined, it is discovered that to a remarkable degree, only the even clusters are present for sizes greater than roughly 40 atoms. These even C_n clusters all turn out to be fullerenes, and the distribution extends out to well past 600 atoms in size (*13, 14*). Of these clusters, as is now well known, some are specially stable chemically—particularly C_{60}, C_{70}, and to a lesser extent C_{84}, C_{76}, and some selected higher fullerenes (*15*). This stability pattern has been

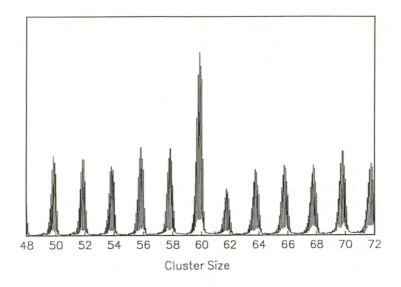

Cluster Size

Figure 1. Boron-doped fullerenes, ($@C_nB_m$), produced by laser vaporization of a boron–graphite composite target disc, and monitored by FTICR mass spectroscopy of the positive cluster ions levitated in a magnetic trap. (Reproduced from reference 9. Copyright 1991 American Chemical Society.)

convincingly explained as a combined result of these molecules' particular ability to adopt smoothly curved structures where no carbon atom has a disproportionate amount of bond-angle strain (and sp^3 character), and the quantum mechanics of forming a closed-shell electronic structure from an edgeless delocalized network. By symmetry, C_{60} in the ($@C_{60}$) soccer-ball structure is the smoothest, roundest possible molecule. It also turns out to be closed-shell, with the largest HOMO–LUMO (highest occupied molecular orbital–lowest unoccupied molecular orbital) gap of any of the fullerenes (*16*).

To incorporate some other heteroatom as part of the cage, bond strength, valency, and size need to be considered. Two elements, boron and nitrogen, are nearly perfect replacements for carbon in the fullerene cage by all these criteria. Accordingly, when boron powder is mixed at a level of 15% by weight with graphite powder, and the composite is pressed into a dense pellet, laser vaporization produced B-doped fullerenes (*9*).

Figure 1 shows the sort of mass spectral evidence that has been obtained for these species. This is a Fourier transform ion cyclotron resonance (FTICR) mass spectrum of the positively charged clusters recorded as they are levitated in the magnetic field of a superconducting magnet. Only clusters with an even number of atoms appear to be present in abundance, and there is extensive substructure within the mass spectrum for each cluster size. Careful analysis of this structure, for the 60-atom clusters, for example, reveals that it results from a mixture of boron-doped fullerenes, ($@C_{60-x}B_x$), where x ranges between 0 and at least 6. For the particular doping experiment probed in Figure 1, the approximate amounts of the various 60-atom clusters were measured to be 22%

C_{60}, 21% $C_{59}B$, 24% $C_{58}B_2$, 18% $C_{57}B_3$, 9% $C_{56}B_4$, 4% $C_{55}B_5$, and 2% $C_{54}B_6$. Similar doping compositions were measured for the other clusters in this 50- to 70-atom size range. The fact that only even-numbered clusters are present here together with the relative importance of the 60-atom clusters suggests that all these species are, in fact, fullerenes. But, aside from the pattern of their masses, how can we tell this for sure? The answer involves probes of their surface chemistry and measures of their photochemistry.

The Cluster FTICR Apparatus. Figure 1 is a mass spectrum, but the apparatus used to generate this spectrum can be used for considerably more than simply examining the masses of a distribution of clusters. Most of the evidence that currently exists for successful doping of fullerenes comes from these more elaborate experiments, and so it is useful to consider this apparatus in more detail.

My colleagues and I have been developing this cluster FTICR apparatus for nearly a decade now as one of the most versatile and powerful means of studying the chemistry and physics of clusters (*17*). Its latest incarnation has been described in detail in a number of publications (*17–19*). A schematic cross section is shown in Figure 2. Briefly, it consists of a pulsed supersonic nozzle−laser-vaporization cluster beam source (*17, 20*) mounted so that the supersonic cluster beam is aimed directly down the central axis of a superconducting magnet. Cluster ions are produced either directly as residual ions from the laser-vaporization event itself within the supersonic nozzle, or as a result of photoionization of the neutral cluster beam just before it passes into the bore of the magnet. Because they are entrained in an intense supersonic beam of helium, all these cluster ions are traveling at speeds close to the terminal beam velocity of 1.9×10^5 cm/s. The cluster's translational energy going into the magnet is then linearly dependent on their mass. For example, a 1000-amu cluster at this velocity has a translational energy of roughly 19 eV, and a 2000-amu cluster moves into the magnet with a 38-eV energy. This supersonic cluster ion beam is mounted so close to the central axis of the magnetic field that the $v \times B$ components of the Lorentz force that are responsible for the magnetic mirror effect turn out to be negligible, and the cluster ions travel smoothly along the periaxial field lines into the center of the magnet.

As shown in the schematic of Figure 2, the cluster ion beam passes through a "deceleration tube" on its way into the magnet. This tube is pulsed under control of a computer to a negative voltage so as to slow down the clusters as they approach the ICR analysis cell. By adjusting the magnitude of this decelerating voltage, one can control the size range of the clusters that will be trapped in the ICR cell.

This analysis cell of the FTICR cell is a 15 cm long with a 4.8-cm-diameter cylinder, fitted with separate front and rear "door" electrodes that are normally kept at a small positive potential in order to trap the cluster ions. In order to inject new clusters into the cell, the front door is typically dropped to 2 V, the rear door is held at 10 V, and the "screen door" shown in the figure is dropped to 0 V. Cluster ions exiting the decelerator tube with translational energies slowed to between 2 and 10 V then pass into the cell and are reflected

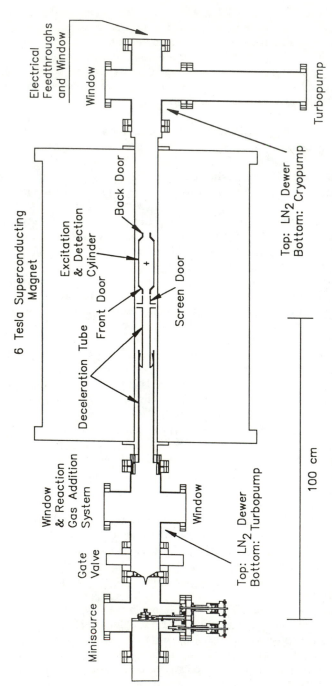

Figure 2. Schematic view of FTICR apparatus used in used to study cluster ions. The clusters are produced via laser vaporization within the small pulsed supersonic beam source in the left-most chamber. The cluster ion beam is then decelerated and ultimately trapped in the ICR analysis cell (the Excitation and Detection Cylinder) mounted in the bore of a 6-T super-conducting magnet. (Reproduced with permission from reference 17. Copyright 1990.)

back from the rear-door electrode. Under computer control the screen door is pulsed back up to 10 V before these bounced cluster ions can escape, thereby trapping them in the cell. Collisions with a low-pressure helium or argon thermalizing gas then remove the 2–10-eV translational energy of these trapped clusters so that they cannot escape when the screen door is dropped to 0 V to let in the next batch of clusters. The result is that cluster ions can be accumulated in the ICR cell until there are enough to study.

Cluster ions are detected in the ICR cell by coherently exciting their cyclotron motion with a computer-crafted rf (radio frequency) waveform, and then monitoring the time-varying tiny image currents induced in the sides of the cell as this cyclotron motion continues for many thousands of cycles. Fourier transformation of this time-domain signal then shows sharp intensity spikes in the frequency domain at the cyclotron frequencies of the cluster ions in the cell—resulting in extremely high resolution, broad-range mass analysis of the contents of the cell.

A critical feature of this apparatus is the fact that the trapping front, rear, and screen door electrodes have large holes (2-cm diameter) through their centers. This feature permits extremely intense laser beams to be directed through the cell to excite or fragment the trapped clusters without excessive scattered light hitting the trapping electrode surfaces. It also permits rapid introduction of reactant gases, and their subsequent evacuation. The result is that any cluster ion that can be made in the supersonic beam source can be probed both for its surface chemistry and for its detailed photophysics and photochemistry while it is levitated in the apparatus, and this can all be probed at a mass resolution between 10^4 and 10^5 to 1. In the discussion to follow, these capabilities of the ICR apparatus are used extensively.

Ammonia Titration of Boronated Fullerenes ($@C_nB_x$).

If, in fact, the laser vaporization of the boron–graphite composite disc did produce true boron-doped fullerenes, the clusters detected in the mass spectrum of Figure 1 should show a very distinct chemistry. Substituting a boron for a carbon atom on a fullerene cage will produce an electron-deficient site at the boron position on the cage. This site should behave as a Lewis acid, and, perhaps even be titrated by moderate pressures of ammonia gas.

Figure 3 shows the result of such a titration experiment on the 60-atom B-doped clusters. In this experiment, the same cluster formation and injection sequence was followed as that used to generate the spectrum of Figure 1, but all clusters except those with 60 atoms were selectively ejected from the cell. This selective ejection was accomplished by a specially crafted rf waveform technology (*18*) known in the trade as SWIFT (stored waveform inverse Fourier transform). The top panel of Figure 3 shows how effective this ejection was in removing all other clusters from the analysis cell. The bottom panel shows the result of exposure to ammonia gas at 1×10^{-6} torr (133 $\times$ 10^{-1} Pa) for 2 s. This is enough time and ammonia gas pressure for the average cluster to have received roughly 300 collisions with ammonia molecules.

The cluster FTICR mass spectrum separates into distinct clumps that, as labeled in the figure, correspond to increasing numbers of chemisorbed

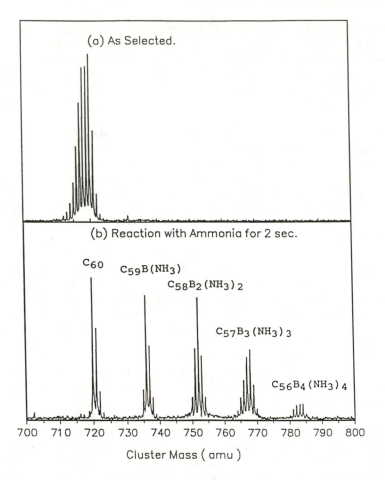

Figure 3. Ammonia titration study of boronated fullerene ions levitated in the magnetic trap of a cluster FTICR apparatus. Clusters of 60-atom size were selected by a SWIFT waveform. Top panel: before reaction; bottom panel: after reaction with 1 × 10^{-6} torr of ammonia for 2 s (roughly 300 collisions). (Reproduced from reference 9. Copyright 1991 American Chemical Society.)

ammonia molecules. The key information here is found in the detailed substructure of each clump in the mass spectrum. The lowest mass clump peaked at 720 amu has the isotopic fine structure expected for pure ($@C_{60}$) with ^{13}C at the natural abundance of 1.1%. The next clump turns out to be just the mass spectral pattern expected for ($@C_{59}B)NH_3$, taking account of the two boron isotopes (20% ^{10}B, 80% ^{11}B). Similar analyses of the mass fine structure of the subsequent clumps in the spectrum show them to be in accord with fully titrated fullerenes with successively higher doping levels: ($@C_{60-x}B_x)(NH_3)_x$.

Further confirmation of the notion that these are really boron-doped 60-atom fullerenes comes from the fact that XeCl excimer laser irradiation of these ammonia-titrated species simply results in the desorption of the ammonia, regenerating the bare 60-atom cluster with an FTICR mass spectrum identical to that obtained, as in the top panel of Figure 3, prior to exposure to ammonia. The attachment of the boron atoms to the cluster is therefore much stronger than the attachment of the chemisorbed ammonia molecules.

The fact that all borons in these fullerene cages appear to be available for titration with ammonia molecules simultaneously indicates that they do not occupy adjacent positions in the cage network. The steric repulsion between neighboring ammonia molecules would otherwise be too great. In addition, the extent of boron substitution into the cage appears to be roughly linear with the size of the fullerene. Combined with the titration result, this linearity is evidence that boron incorporation in the clusters is a random process, little affected by number or positions of previously incorporated boron atoms in the cage.

Laser-Shrinking the Borofullerenes.

One of the most impressive aspects of C_{60}, buckminsterfullerene, is its resistance to fragmentation. Isolated in a vacuum, the positive ion, $(@C_{60})^+$, is the most photophysically stable molecule or cluster we have ever encountered. We estimate that at least 25 eV must be pumped into the molecule before its rate of unimolecular fragmentation becomes fast enough to compete with cooling by infrared emission, and at least 40 eV before it fragments on a microsecond time scale (*21, 22*). Perhaps even more surprising is the fact that the fragmentation that ultimately does occur at these high energies is loss of even-numbered carbon chains, and particularly C_2. This loss is remarkable because C_2 is actually quite weakly bound compared to any other possible molecular carbon fragment. Carbon clusters with 30 atoms or fewer, for example, uniformly favor the C_3 loss channel—a result that makes perfect sense considering how stable the C_3 fragment is.

This weird photofragmentation behavior of C_{60} was found to apply all the other fullerenes as well. Although less resistant to photofragmentation than C_{60}, all other fullerenes were found to fragment by loss of C_2. All, that is, except C_{32}, which seems to disintegrate into many small carbon chain fragments.

When we first observed this behavior in 1987, we postulated (*21*) that it was a result of the fullerenes being closed, edgeless three-connected networks that had no fragmentation channels that did not involve breaking many bonds. For such fullerene network cages, the lowest energy fragmentation pathway we could find involved rearrangement of the pentagon positions on the cage so that two pentagons could share an edge. Such a fullerene can then undergo a concerted C_2 loss where the two carbons connected along the common side of the two pentagons break away and leave as new bonds are formed in the network. The result is that the two original abutting pentagons become one hexagon, and two nearby hexagons are transformed into pentagons. The result is not so much a fragmentation as it is an evaporation: C_2 evaporates off the surface of the cage as the cage shrinks to form the next smaller fullerene.

Improbable as this seems, it is still today the lowest energy fragmentation mechanism yet proposed for a fullerene. It involves a forbidden intramolecular rearrangement in the Woodward–Hoffman sense, but for fullerenes at the fragmentation threshold it is "the only game in town". Given

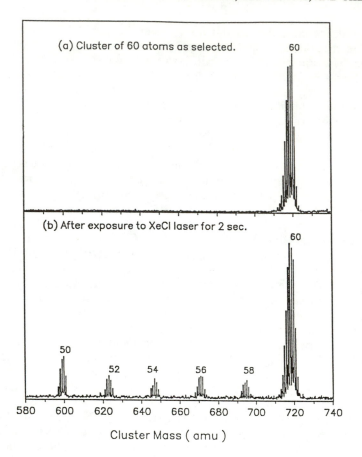

Figure 4. Laser "shrinking" experiment on boron-doped fullerene clusters levitated in FTICR apparatus. The top panel shows the initially selected 60-atom cluster ions. The bottom panel shows the FTICR mass spectrum after irradiation with 100 pulses of a XeCl excimer laser, 308 nm, at 20 mJ/cm² per pulse. (Reproduced from reference 9. Copyright 1991 American Chemical Society.)

sufficiently intense laser excitation, high enough temperatures can be attained so that even these high-energy, forbidden rearrangements occur, and the fullerenes begin to shrink.

Laser-shrinking experiments should therefore provide quite a stringent test of whether the boron-doped carbon clusters seen in Figure 1 really are fullerenes. Under high levels of laser excitation, will they resist fragmentation as well as pure, undoped C_{60}? When they do start to fragment, will they lose C_2? Is there ever a point where the boron "impurity" in the fullerene lattice is preferentially ejected?

Figure 4 shows the result of one such photofragmentation experiment on the SWIFT-selected 60-atom clusters. As seen in the top panel, all clusters other than those of 60 atoms have been ejected, and the complicated substructure in the mass spectrum reveals extensive boron-doping for many of the clusters. The bottom panel shows how this mass spectrum is changed after 100 blasts with a XeCl excimer laser (308 nm) at a fluence of 20 mJ/cm^2 per pulse. Even a cursory examination of this data shows that the boron-doped clusters are behaving very much like pure ($@C_{60}$) under these conditions—just C_2 loss, and at a rate basically the same as the undoped cluster.

Detailed analysis of the mass substructure of these photofragmentation products has now been completed all the way down to the 32-atom limit where pure carbon fullerenes finally burst. For the singly doped fullerenes such as ($@C_{59}B$), there is no evidence for any preferential loss of boron at any stage in the fragmentation. They simply lose C_2 successively, until finally at ($@C_{31}B$) they shatter. The doubly doped fullerenes like ($@C_{58}B_2$) are found to preferentially lose boron in the fragmentation, but this loss is evident in the data only after the cluster has shrunk to fewer than 40 atoms. Clusters with more borons are found to be increasingly likely to lose these borons in the shrinking process, but for clusters with fewer than five boron atoms this preference is hard to detect until the cluster become smaller than 50 atoms.

So it appears that boron-doped fullerenes are nearly as stable with respect to fragmentation as the pure carbon fullerenes, at least for low levels of doping. They are therefore quite reasonable targets for scaling up to bulk production.

How about Nitrogen? In our first experiments attempting to dope boron into carbon cages, we used boron nitride powder, hoping that both boron- and nitrogen-doped fullerenes would be produced. However, we detected only ($@C_nB_x$) clusters. We suspect that this result was simply a problem of titration of most of the free N atoms with other N-containing species to produce N_2. If so, it may be possible to switch successfully to a composite target with the nitrogen being introduced in the form of a CN group.

Doping Inside the Cage

The single most bizarre aspect of fullerenes is that they enclose a vacuum. For C_{60} it is a roughly 3.8-Å-diameter spherical hole, big enough to entrap virtually any atom in the periodic table. In 1985, within a few days of the original realization that C_{60} may be a soccer-ball structure, the first indication was obtained that a metal atom may have been put inside.

This was actually the first example of a doped fullerene ever produced. It was done by laser vaporization of a graphite disc impregnated with $LaCl_3$ (simply by boiling the disc in a water solution of $LaCl_3$ and scraping off the excess $LaCl_3$ crystals from the surface). The result was a complicated mass spectrum of nondescript clusters when they were probed with low laser fluences in the supersonic beam, but at very high fluences of ArF excimer laser power, a stunningly simple mass spectrum was seen. It consisted of just two types of clusters: the normal empty fullerenes C_{60} and C_{70}, and an intense distribution of even-numbered carbon clusters with one (and only one) lanthanum atom attached. At the high-excimer laser fluences being used, the lanthanum atom would have had to be strongly bound, so it was postulated that these were in fact fullerenes with a single lanthanum atom trapped inside.

To use the new fullerene symbol convention introduced here, the new species were thought to have the formula $(La@C_n)$, where strong signal was recorded for all even n from $n = 44$ to >80, and the clusters corresponding to $(La@C_{60})$ and $(La@C_{70})$ were particularly strong.

At the time, this proposal (as well as the fullerene proposal itself for C_{60} and the other even-numbered clusters) generated a bit of controversy. The group at Exxon was particularly skeptical (23), in part because, when trying to replicate our results at low laser fluence with an F_2 excimer laser (7.9-eV photon energy), they found that only a very small fraction of the carbon clusters had a lanthanum atom attached in the neutral cluster beam. Finding that their earlier measurements (24) were incorrect, and that the photoionizations of C_{60} and the other even-numbered carbon clusters were actually two-photon processes with the ArF excimer laser, they argued that the dominance of LaC_n clusters seen when using this laser was due to selective photoionization of these species having low ionization potential. In addition they were able to show that it was possible when using KCl-impregnated graphite targets to attach more than one potassium atom to the C_n clusters (25), raising doubt that there was only one uniquely strong binding site for a metal atom.

Final, unambiguous demonstration that metal atoms actually could be trapped inside the fullerene clusters had to wait a few years until the cluster FTICR apparatus could probe these clusters (26). Carbon clusters with attached metal atoms were generated by laser vaporization, and the positive cluster ions were trapped in the ICR cell. Although several atoms of some metals such as potassium could be attached to the clusters while they were in the free-flying supersonic cluster beam, only the clusters complexed with a single metal atom were sufficiently stable to survive the thermalization collisions encountered with an argon buffer gas as they were trapped in the ICR cell. These MC_n^+ clusters were found to be inert toward reactive gases such as O_2, NH_3, and water, a result suggesting that the metal atom actually was trapped inside and protected by the carbon cage.

But most impressive was the way these MC_n^+ clusters responded to intense laser irradiation. They were found to be just as resistant to fragmentation as the fullerenes, $(@C_n)^+$, and when ultimately they did fragment, they did so by the sequential election of C_2 units. In fact, they behaved in every way just as one would expect for $(M@C_n)^+$ cluster ions, where the C_n fullerene cages were gradually shrink-wrapping around the internal M atom as C_2 molecules evaporated off their surfaces. For empty fullerenes $(@C_n)$, this process was known to terminate abruptly at $(@C_{32})$ where further laser blasting simply resulted in disintegration of the fullerene cage. Now with a metal atom trapped inside, the fullerene cage was found to burst before it could shrink to 32 atoms. In the $(K@C_n)^+$ cluster ions, this minimum survivable cluster size was found to be $(K@C_{44})^+$. Computer modeling revealed this to be the minimum size fullerene cage that can stretch around a central K^+ ion without excessive stressing of the carbon–carbon bonds. The same modeling revealed that the larger ionic radius of Cs^+ would move this bursting point up to fullerene cages 48 or 50 atoms in size. Impressively, the laser "shrink-wrapping" experiments on $(Cs@C_n)^+$ cluster ions did reveal a break-off of the successive C_2 loss mechanism just at this predicted point (26). These experiments provided the most direct test of both the fullerene structural hypothesis, and the notion that true $(M@C_n)$ clusters had been, in fact, been produced simply by laser vaporizing metal-impregnated graphite.

Production of Macroscopic Amounts

Since the fall 1990 announcement by Krätschmer, Huffman, and co-workers (27) of a simple method of producing visible amounts of C_{60} and C_{70}, many groups have tried to extend their technique to the production of doped fullerenes, particularly the trapped-metal species, $(M@C_n)$. In our laboratory at Rice, for example, extensive experiments were tried with graphite rods, drilled out down the center and packed with various metal salts, and metal-salt–graphite composites. When evaporated in the carbon-arc fullerene generator similar to the design (28) that has now become rather standard in the field, the graphitic soots that were produced from these composite graphite rods clearly had metal atoms trapped in some fashion. But extraction with various solvents never produced clear evidence of significant quantities of metal-doped $(M@C_n)$ fullerenes.

Very recently, this situation has changed (11). While considering the possible explanations for why the Krätschmer–Huffman method succeeds in producing C_{60} in such high yield, whereas the laser-vaporization source produced only microscopic quantities, we realized that the key process is probably one of annealing during the cluster growth (28). With this idea in mind, we went back to the use of a laser to vaporize the graphite target, but this time arranged that the vaporization be done in a furnace at 1200 °C. The higher temperature, we suspected, would increase the rate of annealing of the growing graphitic sheets, causing them to curl back on themselves in the optimum manner. This route in the growth kinetics we have dubbed the "pentagon road" because it follows a path where the graphitic sheet tends to have as many pentagons as possible (thereby minimizing dangling bonds), while avoiding the instability and high chemical reactivity occasioned by adjacent pentagons. In fact, we found that this laser vaporization of graphite targets in an oven at 1200 °C readily produces C_{60} and C_{70} at very high yields—well over 10% of all the vaporized carbon.

Just recently laser vaporization of composite target rods made of 10% La_2O_3 mixed with graphite powder in such an oven at 1200 °C, in the presence of a slow flow of argon at 250 torr (33.3×10^3 Pa), was found to produce $(La@C_n)$ clusters in milligram quantities (11). The initial result of this laser vaporization is a black–brown deposit that collects on the cool downstream end of the surrounding quartz tube as it passes out the end of the 1200-°C furnace. Subsequent heating of this deposit to 650 °C results in sublimation of a

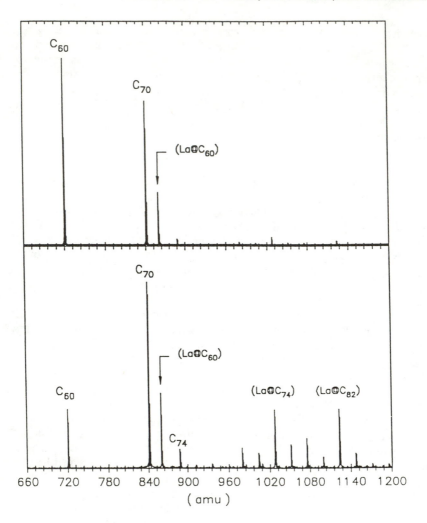

Figure 5. FTICR mass spectrum of black, mirror-like sublimed film showing presence of (La@Cn) fullerenes. For the top panel the pulsed decelerator responsible for slowing the clusters down to be trapped in the analysis cell was optimized for the C_{60}–C_{70} mass region. For the bottom panel it was optimized for the region around C_{84}. Reaction studies with O_2 and NH_3 revealed all these clusters to be chemically inert. Laser "shrink-wrapping" studies proved the them to be fullerenes. (Reproduced from reference 11. Copyright 1991 American Chemical Society.)

material that deposits as a black, nearly mirror-quality film on nearby cold surfaces. Gentle laser probing of this film (which is stable in air) reveals that it is largely made up of a matrix of (@C_{60}) and (@C_{70}) in which lanthanum fullerenes, (La@C_n) are safely entrapped. As shown in Figure 5, the dominant metal-containing species are (La@C_{60}), (La@C_{70}), (La@C_{74}), and (La@C_{82}).

If, instead of sublimation to form a film, the raw graphitic soot from laser vaporization at 1200 °C is collected and subjected to a Soxhlet extraction for several hours with toluene, a solution of fullerenes is obtained, and probing by gentle laser desorption shows that one particular lanthanum fullerene, ($La@C_{82}$), is sufficiently soluble and chemically stable to survive boiling in toluene and exposure to air. It appears to be the first macroscopic sample of metal-doped fullerenes yet produced.

The fact that this stable metal fullerene has 82 carbon atoms at first seems strange. Why not the 60-atom species ($La@C_{60}$) seen in Figure 5 to be most abundant in the sublimed film? The answer, however, is fairly obvious. The most stable empty fullerene, ($@C_{60}$), has a closed-shell electronic structure with a very large HOMO–LUMO gap. When a metal atom like lanthanum is inserted into the inner cavity, some electrons will be transferred from the metal to the carbon cage. In ($La@C_{60}$), Chang et al. calculate (29) that the ground electronic state will be an electronically open-shell molecule with a spin of 3/2. The dominant electronic configuration is one where the two 6s electrons of the lanthanum atom have been donated to the carbon cage t_{1u} LUMO orbital, leaving behind a single unpaired electron in the La 5d orbital. Although this molecule will be quite stable while isolated in the gas phase, it will likely be rather reactive in condensed phases. Such higher reactivity can be expected to be a general feature of doped fullerenes.

In ($La@C_{82}$), however, donation of two electrons from the lanthanum produces a delocalized "pi" shell electron count of 84. Although confirming calculations have yet to be done, it is expected that this π shell electron count will produce a stable closed-shell electronic state with a substantial HOMO–LUMO gap, much as occurs with ($@C_{84}$), which is the third-most stable empty fullerene, after ($@C_{60}$) and ($@C_{70}$). Corresponding 60- and 70-shell electron lanthanum fullerenes will occur for ($La@C_{58}$) and ($La@C_{68}$), but these fullerene cages have adjacent pentagons and will therefore be highly reactive.

Extensive research is now underway to examine the properties of this stable ($La@C_{82}$) metal-doped fullerene and to find ways of passivating the more reactive, open-shell species like ($La@C_{60}$). The fact that it can be produced initially in an air-stable, sublimed film of ($@C_{60}$) and other stable fullerenes provides a convenient means of transporting them to an air- and water-free environment for detailed chemical workup.

Preliminary evidence has also been obtained from this laser-vaporization apparatus for the formation of sublimed fullerene films containing potassium-doped fullerenes, ($K@C_n$), boron-doped fullerenes like ($@C_{59}B$), and mixed boron–potassium doped species like ($K@C_{59}B$). The production and study of a broad range of doped fullerenes may soon be possible. Furthermore, once the chemical properties of these species are known, it may be possible to go back to carbon-arc-based methods or others that can be extended to large-scale production. In fact, we have obtained evidence already that the ($La@C_{82}$) species can be extracted from carbon-arc-produced soot (11).

Doping Preformed Fullerenes

Instead of the aforementioned methods of producing doped fullerenes, all of which in a sense "start from scratch", an alternative would be to begin with pure fullerenes such as C_{60} or C_{70} as a starting material. Several groups have recently shown that it is possible to insert a helium atom inside a C_{60} cage, forming (He@C_{60}) simply by ion bombardment (7, 8, 30, 31). Although these experiments were done only a few molecules at a time in a mass spectrometer by firing a $(C_{60})^+$ ion beam at 4–10 keV through a short chamber filled with helium gas, they raise the interesting possibility that doped fullerenes might be produced effectively by an extension of the ion-implantation techniques currently used to dope semiconductor films.

Imagine, for example, a device that deposits C_{60} on a metal substrate by sublimation of a pure C_{60} sample in an oven while simultaneously bombarding this film with lithium ions at energies in the 10–100-eV range. The Li^+ ion is considerably smaller than He, so it is reasonable to expect that a this lithium ion implantation device might produce a substantial number of (Li@C_{60}) species as the C_{60} and Li^+ ions are deposited together on the metal substrate. Of course there will be considerable fragmentation of some of the C_{60} cages to produce smaller (Li@C_n) fullerenes that, because they have adjacent pentagons, will be rather unstable; and many of the lithium atoms will end up outside the fullerene cages. But a subsequent workup of this film by sublimation or dissolution in appropriate solvents may permit the (Li@C_{60}) material to be isolated in usable amounts.

In view of the likeliness of a small amount of fragmentation accompanying this direct ion-implantation process, it may be best to start with C_{70} as the initial fullerene target. This initially impregnated (Li@C_{70}) "egg" would then get rid of its excess energy, in part, by fragmenting (shrink-wrapping) down to (Li@C_{60}). Although pure C_{70} is currently a very expensive starting material, it is made in considerable abundance in carbon arcs (32) and sooting flames (33), and should ultimately be a rather inexpensive commodity.

In addition to such brute-force methods of producing metal-filled fullerenes, there will almost certainly be effective chemical routes developed to achieve these same ends. Starting with pure C_{60}, C_{70}, or larger fullerenes, it seems reasonable to hope that future fullerene chemists will be able to open up the cage in a controlled manner, cut it down by a desired amount, and reassemble these cage parts again, trapping desired atoms or molecules in the inside. As a side benefit, this fullerene chemistry of the future may permit the synthesis of a wide range of interconnected fullerene cage structures or even hollow graphitic tubes and fibers.

Conclusions

Now that it is clear that some of these species can be made in at least small amounts, the future of research in this "dopy" fullerene field seems exceedingly bright. Once the chemical and physical properties of a few of these species are understood, the way will be open for a much broader and more systematic study. In the long run there will probably be many ways found to fill the fullerene cages, substitute for one or more of the cage carbon atoms, and assemble these objects together to form useful materials. Conceivably, it will be possible to have the entire periodic table done again, this time with an atom of each element, E, encapsulated in a fullerene cage. These (E@C$_n$) superatoms may prove to be the building blocks of a vast array of new nanometer-engineered materials. But first, the newly emerging fields of fullerene chemistry, physics, and materials science have some major homework to do.

Acknowledgments

The research described here from my group at Rice has resulted from wonderful collaborations with an extended group of superb students, postdoctoral associates, and more senior colleagues over the past decade. My long-time partner in cluster research, Ori Cheshnovsky of the University of Tel Aviv, is responsible for the charming suggestion that @ be used in specifying complex fullerenes. Support for this work has come from the National Science Foundation, the Office of Naval Research, and the Robert A. Welch Foundation. The cluster FTICR apparatus, so crucial to these studies, was developed with major support from the U.S. Department of Energy, Division of Chemical Sciences.

References

1. Saito, S.; Oshiyama, A. *Phys. Rev. Lett.* **1991,** *66,* 2637.

2. Martins, J. L.; Troullier, N.; Weaver, J. H. *Chem. Phys. Lett.* **1991,** *180,* 457.

3. Skumanich, A. *Chem. Phys. Lett.* **1991,** *182,* 486.

4. Miller, B.; Rosamilia, J. M.; Dabbagh, G.; Tycko. R.; Haddon, R. C.; Muller, J.; Wilson, W.; Murphy, D. W.; Hebard, A. *J. Am. Chem. Soc.* **1991,** *113,* 6291.

5. Haddon, R. C.; Hebard, A.; Rosseinsky, M. J.; Murphy, D. W.; Duclos, S. J.; Lyons, K. B.; Miller, B.; Rosamilia, J. M.; Fleming, R. M.; Kortan, A. R.; Glarum, S. H.; Makhija, A. V.; Muller, A. J.; Eick, R. H.; Zakurak, S. M.; Tycko, R.; Dabbagh, G.; Thiel, F. A. *Nature (London)* **1991,** *350,* 320.

6. Heath, J. R.; O'Brien, S. C.; Zhang, Q.; Liu, Y.; Curl, R. F.; Kroto, H. W.; and Smalley, R. E. *J. Am. Chem. Soc.* **1985,** *107,* 7779.

7. Weiske, T.; Bohme, D. K.; Hrusak, J.; Krätschmer, W.; Schwarz, H. *Angew. Chem. Int. Ed. Engl.* **1991,** *30,* 884.

8. Ross, M. M.; Callahan, J. H. *J. Phys. Chem.* **1991,** *95,* 5720.

9. Guo, T.; Jin, C.; Smalley, R. E., *J. Phys. Chem.* **1991,** *95,* 4980.

10. Tanigaki, K., Eddesen, T. W.; Saito, S.; Mizuki, J.; Tsai, J. S.; Kubo, Y.; Kuroshima, S. *Nature (London)* **1991,** *352,* 222.

11. Chai, J.; Guo, T.; Jin, C,; Haufler, R. E.; Chibante, L. P. F.; Wang, L.; Alford, J. M.; Smalley, R. E. *J. Phys. Chem.* **1991,** *95,* 7564–7568.

12. Kroto, H. W; Heath, J. R.; O'Brien, S. C.; Curl, R. F.; Smalley, R. E.; *Nature (London)* **1985,** *318,* 162.

13. So, H. Y.; Wilkens, C. L. *J. Phys. Chem.* **1989,** *93,* 1184.

14. Maruyama, S.; Lee, M. Y.; Haufler, R. E.; Chai, Y.; Smalley, R. E. *Z. Physik D.* **1991,** *19,* 409.

15. Diederich, F.; Ettl, R.; Rubin, Y.; Whetten, R. L.; Beck, R.; Alvarez, M.; Anz, S.; Sensharma, D.; Wudl, F.; Khemani, K. C.; Koch, A. *Science (Washington, D.C.)* **1991,** *252,* 548.

16. Yang, S. H.; Taylor, K. J.; Craycraft, M. J.; Conceicao, J.; Pettiette, C. L.; Cheshnovsky, O.; Smalley, R. E. *Chem. Phys. Lett.* **1988,** *144,* 431.

17. Maruyama, S.; Anderson, L. R.; Smalley, R. E. *Rev. Sci. Instrum.* **1990,** *61,* 3686–3693.

18. Alford, J. M.; Laaksonen, R. T.; Smalley, R. E.; *J. Chem. Phys.* **1991,** *94,* 2618.

19. Maruyama, S.; Anderson, L. R.; Smalley, R. E. *J. Chem. Phys.* **1990,** *93,* 5349.

20. Dietz, T. G.; Duncan, M. A.; Powers, D. E.; Smalley, R. E. *J. Chem. Phys.* **1981,** *74,* 6511.

21. O'Brien, S. C.; Heath, J. R.; Curl, R. F.; Smalley, R. E. *J. Chem. Phys.* **1988,** *88,* 220.

22. Radi, P. P.; Bunn, T. L.; Kemper, P. R.; Moichan, M. E.; Bowers, M. T. *J. Chem. Phys.* **1988,** *88,* 2809.

23. Cox, D. M.; Trevor, D. J.; Reichmann, K. C.; Kaldor, A. *J. Am. Chem. Soc.* **1986,** *108,* 2457.

24. Rohlfing, E. A.; Cox, D. M.; Kaldor, A.; *J. Chem. Phys.* **1984,** *81* 3322.

25. Cox, D. M.; Reichmann, K. C.; Kaldor, A. *J. Chem. Phys.* **1988,** *88,* 1588.

26. Weiss, F. D.; O'Brien, S. C.; Elkind, J. L.; Curl, R. F.; Smalley, R. E., *J. Am. Chem. Soc.* **1988,** *110,* 4464.

27. Krätschmer, W.; Lamb, L. D.; Fostiropoulos, K.; Huffman, D. R. *Nature (London)* **1990,** *347,* 354.

28. Haufler, R. E.; Chai, Y.; Chibante, L. P. F.; Conceicao, J.; Jin, C.; Wang, L-S.; Maruyama, S.; Smalley, R. E. *Mater. Res. Soc. Symp. Proc.* **1990,** *206,* 627.

29. Chang, A. H. H.; Ermler, W. C.; Pitzer, R. M. *J. Chem. Phys.* **1991,** *94,* 5004.

30. Weiske, T.; Hrusak, J.; Bohme, D. K.; Schwarz, H. *J. Am. Chem. Soc.* **1991,** in press.

31. Caldwell, K. A.; Giblin, D. E.; Hsu, C. S.; Cox, D. M.; Gross, M. L. *J. Am. Chem. Soc.* **1991,** in press.

32. Parker, D. H.; Wurz, P.; Chatterjee, K.; Lykke, K. R.; Hunt, J. E.; Pellin, M. J.; Hemminger, J. C.; Gruen, D. M.; Stock, L. M. *J. Am. Chem. Soc.* **1991,** *113,* 7499.

33. Howard, J. B.; McKinnon, J. T.; Makarovsky, Y.; Lafleur, A.; Johnson, M. E. *Nature (London)* **1991,** *352,* 139.

Received September 4, 1991

Chapter 11

Survey of Chemical Reactivity of C_{60}, Electrophile and Dieno–polarophile Par Excellence

F. Wudl, A. Hirsch, K. C. Khemani, T. Suzuki, P.-M. Allemand, A. Koch, H. Eckert, G. Srdanov, and H. M. Webb

Institute for Polymers and Organic Solids and Departments of Physics and Chemistry, University of California, Santa Barbara, CA 93106

Buckminsterfullerene, C_{60}, was found to add a large variety of neutral and charged nucleophiles, a couple of dienes, and a number of dipoles. The reagents mentioned within are propylamine, dodecylamine, tert-butylamine, ethylenediamine, morpholine, triethyl phosphite, hydride (lithium triethyl borohydride), phenyllithium, phenylmagnesium bromide, tert-butyllithium, tert-butylmagnesium bromide, anthracene, cyclopentadiene, p-nitrophenyl-azide, ethyl diazoacetate, phenyldiazomethane, diphenyldiazomethane, and the ylide $(CH_3)_2S^+CH^-CO_2Et$. When a large molar excess of reagent relative to fullerene is used, as many as 10–12 new bonds are formed. For primary amine addition (with the exception of tert-butylamine), the mechanism involves single-electron transfer as a first step. Mass spectrometry shows facile reactivity of C_{60} in the gas phase with CH_4 in both positive and negative ionization.

A large number of papers on the characterization of fullerenes have appeared (*1–16*), but very few publications describe their chemical reactivity (*17–20*). As has been known for some time, C_{60} and C_{70} are mild oxidizing agents (*6, 12*); conversely, C_{60} and C_{70} are rather poor reducing agents. The position of the first reduction wave indicates that, in its reactivity, $I_2 > C_{60}$, $C_{70} > BQ$ (BQ is benzoquinone). The relatively high electronegativity of this carbon allotrope was proposed (*12*) to be due to the pyracylene character of certain inter-five-membered ring bonds. In view of that observation, we hypothesized that C_{60} would be readily attacked by nucleophiles and would be a good dienophile, akin to benzoquinone.

In this chapter, we disclose a survey of the reactivity of C_{60} with neutral and charged nucleophiles as well as dienes and a dipole. Because C_{60} is a multifunctional molecule (it has six interconnected pyracylene moieties), most

0097–6156/92/0481–0161$06.00/0

attempts to perform stoichiometric reactions are expected to produce mixtures of difficult-to-separate products. Also, the available quantities of C_{60} are still limited, and only tens of milligrams per reaction could be employed. We therefore decided to test the hypothesis by use of large excess of reagent relative to C_{60}. Preliminary results indicated that acid quenching of the addition products of certain nucleophiles resulted in reversal with the concomitant formation of C_{60}; quenching the reaction mixtures with methyl iodide prevents the problem (21).

Results and Discussion

As can be seen from the results in the equations in Scheme I, C_{60} adds as many as 10 carbon anions, presented either in the form of Grignard reagents or lithium reagents. We believe the difference in results between commercial phenyl Grignard and that prepared in situ from phenyllithium is due to the presence of colloidal magnesium metal in the commercial reagent. (We thank Professor George A. Olah of the University of Southern California for the suggestion that colloidal Mg may be responsible for this strange reactivity). The magnesium, in competition with phenyl addition, acts as a reducing agent, and the reduced species is quenched with methyl iodide to produce material in which the ratios of Me to Ph are >1. As mentioned later and shown in Figure 1, nucleophilic addition is not restricted to the condensed phase because the elements of negatively charged methane add to fullerene C_{60} in the gas phase.

The only nucleophilic reagent with which C_{60} formed a stoichiometric adduct was $LiB(Et)_3H$. The latter was possible by the discovery of immediate precipitation of a C_{60} lithium salt from benzene upon addition of a THF solution of the borohydride. This hydrofulleride salt ($C_{60}HLi \cdot 9H_2O$) does not react with methyl iodide and is indefinitely stable in methanol solution at room temperature, results implying that its conjugate acid $C_{60}H_2$ has a pK_a << 16. In accord with the redox properties, iodide and thiocyanate, so far, were found to be unreactive toward C_{60}. (Because I_2 and $(SCN)_2$ are better oxidizing agents than C_{60}, I^- and SCN^- are more stable than C_{60}^-.)

The most remarkable result is the addition of neutral nucleophiles such as amines. As many as 12 propylamine molecules can be added onto a C_{60}. The ethylenediamine, propylamine, and morpholine adducts are soluble in dilute hydrochloric acid, whereas the dodecylamine adduct is not.

The mechanism of addition is stepwise; in ethylenediamine and propylamine addition, electron transfer precedes covalent bond formation, as determined spectroscopically and by electron spin resonance (ESR) spectroscopy. With both techniques, a signal due to radical anion formation is replaced by that of the adduct. The decay of the open shell intermediate is slow for propylamine (on the order of hours, see Figure 2) and faster for phenyl Grignard

Primary Amine Addition

$$C_{60} + RNH_2 \text{ (solvent)} \xrightarrow{\text{RT, 24 hr}} C_{60}(NH_2R)_x$$

R = Pr, x ≤ 12; R = t-Bu, x = 10

R = $C_{12}H_{25}$, x > 1;

R = $(CH_2)_2(NH_2)_2$, x ~ 6

Electrocyclic Reactions

$$C_{60} + \text{xs Anthracene.} \xrightarrow[120°]{\text{ODCB}} C_{60}(C_{14}H_{14})_x$$

$$C_{60} + \text{CpH (solvent)} \xrightarrow[\text{RT}]{} C_{60}(CpH)_x$$

CpH = cyclopentadiene

$$C_{60} + \text{xs } O_2NC_6H_4N_3 \xrightarrow[110°, 3 \text{ hr}]{C_6H_5Cl} C_{60}(N_3C_6H_4NO_2)_x$$

ODCB = o-dichlorobenzene; xs = excess

Grignard & Organolithium Reagent Addition

$$60 \text{ PhMgBr}^a + C_{60} \xrightarrow{\text{THF, RT}} C_{60}Ph_{10}Me_{10}$$

$$60 \text{ PhLi} + C_{60} \xrightarrow[\text{2) MeI}]{\text{THF, RT}} C_{60}Ph_3Me_5$$

$$60 \text{ PhMgBr}^b + C_{60} \xrightarrow[\text{2) MeI}]{\text{THF, RT}} C_{60}Ph_2Me_3$$

$$60 \text{ } t\text{-BuMgBr}^a + C_{60} \xrightarrow[\text{2) MeI}]{\text{THF, RT}} C_{60}t\text{-Bu}_{10}Me_{10}$$

$$60 \text{ } t\text{-BuMgBr}^b + C_{60} \xrightarrow[\text{2) MeI}]{\text{THF, RT}} C_{60}t\text{-Bu}_{10}Me_{10}$$

$$60 \text{ } t\text{-BuLi} + C_{60} \xrightarrow[\text{2) MeI}]{\text{THF, RT}} C_{60}t\text{-BuMe}$$

[a] From PhLi and MgBr$_2$

[b] Commercial reagent

Scheme I

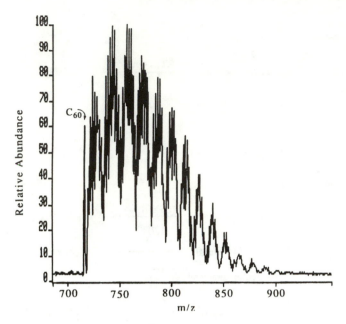

Figure 1. *Negative ion chemical ionization mass spectrum of* C_{60} *using methane as buffer gas.*

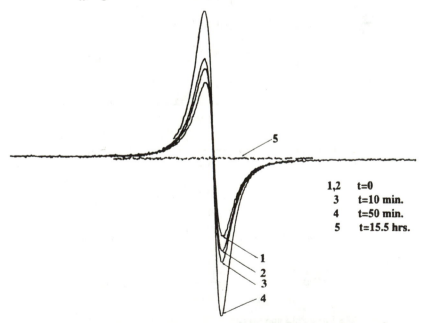

Figure 2. *Electron spin resonance spectroscopy as a function of time of a reaction of* C_{60} *with propylamine in propylamine solvent. Curves 1 and 2 were taken within a minute of each other. The spectra were taken at 298 K; the line width (peak-to-peak) is 1.25 G. After 15 h there was no* C_{60} *left as determined by thin layer chromatography.*

addition (on the order of minutes). With *tert*-butylamine, the reaction occurs only in polar solvents such as dimethylformamide (DMF), and no intermediate radical species is observed, but a deep blue intermediate can be followed spectrophotometrically. Morpholine, an example of a secondary amine, adds smoothly and affords a precipitate that consists of $C_{60}[HN(C_2H_4)_2O]_7$, as determined by fast atom bombardment mass spectrometry (FABMS), but $C_{60}[HN(C_2H_4)_2O]_6$ as determined by elemental analysis. In preliminary work (Hirsch, A., University of California, Santa Barbara, unpublished data), we found that triethyl phosphite adds to C_{60} to produce an Arbutzov-type of product as determined by infrared spectroscopy (alkyl phosphonate bands). As many as six phosphonate groups are added (FABMS).

Negative ion chemical ionization in the mass spectrometer involved vaporization of C_{60} at 250 °C in a bath gas of CH_4 at a pressure of 1 torr (133 Pa); methane acts as an efficient moderating medium to reduce the kinetic energy of high energy (300 eV) electrons for use in resonant electron capture by a substrate. Under these conditions, however, C_{60} exhibited, in addition to simple electron capture, a high degree of reactivity toward CH_4 (Figure 1). This behavior is unusual, as methane rarely appears as a participant in these reactions. Negative ions are observed at every other mass starting with 720 with a local maximum at 724 amu ($C_{60} + 2H_2$). From this point to approximately 900 amu, regular and descending maxima are observed with a mass separation of approximately 14 amu, consistent with successive addition of methylene. The maximum number of carbons attached to C_{60} seems to be 13 on the basis of the number of clearly identifiable mass envelopes. Positive ion chemical ionization of C_{60} also shows much reactivity with methane. In this case, high concentrations of $C_2H_5^+$ plus other methane condensation products can more readily account for the abundance of masses relative to the negative ion spectrum. In both sets of experiments it is clear that, in the gas phase, C–C bonds are being formed with C_{60}.

The dienophile reactivity of the C_{60} double bonds is expressed in the formation of adducts with the mildly reactive dienes anthracene and furan as well as the more reactive cyclopentadiene. As can be seen in Scheme I, *p*-nitrophenyl azide also adds to C_{60}, a result attesting to C_{60}'s dipolarophile nature. Other dipoles, not shown, such as ethyl diazoacetate (Suzuki, T., University of California, Santa Barbara, unpublished data), phenyldiazomethane (Suzuki, T., unpublished data), diphenyldiazomethane (Suzuki, T., unpublished data), and the ylide $(CH_3)_2S^+CH^-CO_2Et$ (Suzuki, T., unpublished data) also give adducts.

The reactions in Scheme I can be envisioned to proceed as shown in Scheme II, where we do not imply that the cycloadditions are concerted or that the nucleophilic additions do not proceed through SET (single-electron-transfer processes).

In view of Scheme II, it is not surprising that C_{60} reacts so effectively with OsO_4, particularly if the tetroxide reacts in an electrocyclic process or is

A. Charged Nucleophile Addition

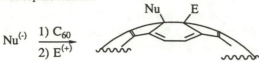

B. Amine Addition

$R\ddot{N}(R')H \xrightarrow{C_{60}}$

R' = R, H

C. Dipolar Cycloaddition

a, b, c = N; a, b, = N, c = CR$_2$

D. 2 + 4 Cycloaddition

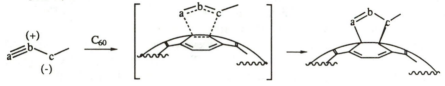

Scheme II

converted to a nucleophilic species by coordination with pyridine (*22*). It is equally not surprising that $(Ph_3P)_3Pt(\eta^2\text{-}C_2H_4)$ undergoes an oxidative addition (*23*) to C_{60}, in a fashion similar to the dipolar cycloaddition reaction depicted in Scheme II.

Experimental Section

Buckminsterfullerene, C_{60}, was prepared in our newly designed bench-top reactor (*24*). Tetrahydrofuran (THF) distilled freshly over sodium metal was used. UV–visible spectra were recorded on a Perkin-Elmer Lambda-5 spectrophotometer. IR spectra were taken with a Perkin-Elmer model 1330 spectrophotometer. ^{1}H NMR and ^{13}C NMR spectra (500 or 300 MHz) were recorded on a General Electric model GE-500 or GN-300 spectrometer; all samples were internally calibrated with tetramethylsilane (TMS). Thermal analyses were carried out on a Perkin-Elmer TGS-2 thermogravimetric analyzer. FAB mass spectra were recorded on a VG 70–250 HF spectrometer (Xenon 7–8 keV; matrix was *m*-nitrobenzyl alcohol).

Purification of C_{60}. Pure C_{60} was obtained by column chromatography. In a typical run about 400 mg of extract consisting of C_{60}, C_{70}, and higher fullerenes was passed through a column of Brockmann neutral activity grade I, 80–200 mesh alumina (1.5 kg). The purple C_{60} was eluted first with 5% toluene in hexanes. The second fraction was the red-brown C_{70} (20% toluene in hexanes). Higher fullerenes can be extracted with pure toluene. The yields are 53% C_{60} by weight from extract and 11% C_{70}, still containing some C_{60} (HPLC, hexanes–silica gel). For further purification C_{60} was dissolved in benzene, precipitated with hexanes, filtered, and dried under vacuum at room temperature.

Formation of Adducts of C_{60} with Primary and Secondary Amines.

$C_{60}(NH_2C_3H_7)_x$. C_{60} (15.5 mg, 2.17 × 10^{-5} mol) was stirred in 20 mL of freshly distilled *n*-propylamine for 24 h at room temperature. A green solution that was formed turned brown after about 0.5 h. The *n*-propylamine was evaporated, and the brown product was dried under vacuum at room temperature (r.t.). Yield: 20.6 mg of $C_{60}(NH_2C_3H_7)_x$. For purification of the crude product, which contains no unreacted C_{60} (TLC, silica gel, toluene–hexane 1:10), crude product was dissolved in a small amount (1 mL) of CHCl$_3$ and precipitated with hexane (20 mL) from the red-brown solution. After centrifugation, the yellow-brown solid was washed with hexane (twice) and dried under vacuum at r.t. for 24 h.

Elemental analyses (found: C, 78.14%; H, 4.76%; N, 6.96%) were difficult to reconcile with rational formulae, but the ratios H:C and H:N, assuming incomplete combustion, indicated that the average number of propylamines added was 6. IR (KBr) ν (wave number): 3300 (m), 2950 (s), 2920 (m), 2870 (m), 1450 (s, br), 1370 (w), 1070 (s, br), 870 (w, br), 530 (w, br), 410 (m) cm^{-1}. FABMS: clusters of peaks at 1370 ($C_{60}(NH_2C_3H_7)_{11}$), three clusters consistent with the loss of three carbons; 1311 ($C_{60}(NH_2C_3H_7)_{10}$, several clusters; 1075 ($C_{60}(NH_2C_3H_7)_6$), three clusters; 1016 ($C_{60}(NH_2C_3H_7)_5$), three clusters; 956 ($C_{60}(NH_2C_3H_7)_4$), three clusters; 897 ($C_{60}(NH_2C_3H_7)_3$), three clusters; 838 ($C_{60}(NH_2C_3H_7)_2$), three clusters; 779 ($C_{60}(NH_2C_3H_7)$), three clusters; 720 (C_{60}) amu. ^{1}H NMR (CDCl$_3$) δ (TMS): 3.8–2.4 (br, 3 H; CH$_2$, CH), 1.9–1.3 (br, 2 H; CH$_2$), 1.25 (br 1 H; NH), 1.3–0.6 (br, 3 H; CH$_3$) ppm. Thermogravimetric analysis (TGA): wt. loss 50–340 °C, 26%.

$C_{60}(NH_2C_{12}H_{25})_x$. C_{60} (20 mg, 2.8×10^{-5} mol) was stirred in 1 mL of commercial dodecylamine for 24 h at 30 °C (the initially formed green suspension turned brown after about 1 h). Then methanol was added, and the brown precipitate was centrifuged and washed (twice) with methanol. For further purification, the crude product, which contains no unreacted C_{60} (TLC), was dissolved in a small amount (1 mL) of $CHCl_3$ and precipitated with hexane (20 mL) from the red-brown solution. After centrifugation the yellow-brown solid was washed with hexane (twice) and dried under vacuum–r.t for 24 h.

IR (KBr) ν: 3300 (vw), 2950 (sh), 2920 (s), 2850 (m), 1460 (s, br), 1320 (w, br), 1110 (s, br), 1030 (s, br), 530 (m, br), 410 (m, br) cm^{-1}. FABMS: cluster of peaks at 905 ($C_{60}(NH_2C_{12}H_{25})$), several clusters of fragmentation peaks, 720 (C_{60}) amu. ^1H NMR ($CDCl_3$) δ: 3.8–2.6 (br, 2 H; CH_2), 2.2–0.6 (br, 25 H; CH, CH_2, NH, CH_3) ppm. ^{13}C NMR ($CDCl_3$) δ: 155–142 (br), 31.96, 29.74, 27.50 (br), 22.71, 14.17 ppm. TGA: wt. loss 50–340 °C, 37%.

$C_{60}(NH_2C_4H_9)_x$. C_{60} (22 mg, 3.2×10^{-5} mol) was stirred in a mixture of 20 mL of commercial t-butylamine and freshly distilled DMF (1:3) for 16 h. The initially formed blue solution turned brown after 1 h. The t-butylamine was evaporated, and the product was precipitated with water and centrifuged. After the liquid was decanted, 5 mL of $CHCl_3$ was added, and insoluble side products were removed by centrifugation. After evaporation of the $CHCl_3$, the yellow-brown solid was dried under vacuum at r.t.

IR (KBr) ν: 3250 (w), 2950 (s), 2920 (m), 2850 (w), 1450 (m, br), 1390 (m), 1360 (m), 1210 (s, br), 1100 (s, v br), 540 (w, br), 410 (w, br) cm^{-1}. FABMS: several clusters of peaks (fragmentations) between 720 and 1400 (corresponding to as many as 10 t-butylamino groups); clusters of peaks at 866 ($C_{60}(NH_2C_4H_9)_2$), 793 ($C_{60}(NH_2C_4H_9)$), 720 (C_{60}) amu. UV–vis (blue intermediate in DMF) λ_{max}: 605, 530 nm. ^1H NMR ($CDCl_3$–TMS) δ: 1.8 (v br, 1 H; CH), 1.7–1.1 (br, 9 H; CH_3), 1.25 (^1H; NH) ppm.

$C_{60}(C_2H_8N_2)_x$. C_{60} (15 mg, 2.2×10^{-5} mol) was stirred in 20 mL of freshly distilled ethylenediamine for 1 day. After 2 h the green solution, which showed a doublet ESR signal, turned brown and became diamagnetic. The product was precipitated with THF, centrifuged, washed (twice, THF) and dried under vacuum at r.t. The crude solid was dissolved in 2 mL of water, and the insoluble material was separated by centrifugation. THF (20 mL) was added to the yellow solution to precipitate the product. After centrifugation the yellow-brown powder was dried under vacuum at r.t.

IR (KBr) ν: 3250 (m), 2980 (w), 2960 (w), 2940 (sh), 1450 (s, br), 1110 (s), 1020 (sh), 860 (w) cm^{-1}. FABMS: clusters of peaks at 781 ($C_{60}(NH_2C_2H_4$ + H), 720 (C_{60}) amu. ^1H NMR (D_2O, D_2SO_4) δ: 3.0 ppm. Titration of 1.4 mg of $C_{60}(NH_2C_2H_4NH_2)_x$ with 0.5 N HCl: titration curve shows two steps corresponding to the two amino groups per attached ethylenediamine unit; 32 μL of HCl is needed for the titration, an amount that is consistent with 12 amino groups per molecule, leading to the average stoichiometry $C_{60}(NH_2C_2H_4)_6$.

C$_{60}$(NHC$_4$H$_8$O)$_x$. C$_{60}$ (20 mg, 2.8 × 10^{-5} mol) was stirred in 5 mL of freshly distilled morpholine at room temperature. The green suspension that was formed first turned into an orange suspension. After 40 h all of the C$_{60}$ reacted (TLC, silica gel; toluene–hexane 1:10). The excess morpholine was removed by distillation under reduced pressure and drying under vacuum at r.t. For further purification, the material was dissolved in 1 mL of CHCl$_3$ and precipitated with 5 mL of hexane from the red-brown solution. The precipitate was centrifuged and washed (twice) with hexane. The orange-brown solid was dried for 24 h under vacuum at r.t.

Anal.: Calcd. for C$_{60}$(NHC$_4$H$_8$O)$_6$ · (CHCl$_3$)$_{0.5}$: C, 78.67; H, 3.77; N, 6.47. Found: C, 78.32; H, 3.84; N, 6.44. IR (KBr) ν: 2970 (m), 2950 (sh), 2920 (m), 2910 (sh), 1450 (sh), 1395 (w), 1290 (w), 1270 (m), 1210 (w), 1120 (s), 1070 (w), 1010 (m), 880 (m), 840 (sh), 550 (w), 530 (w) cm^{-1}. FABMS: clusters of peaks up to 1300 amu, fragmentations between 1300 and 720 (C$_{60}$), and parent peak at 807 (C$_{60}$(NHC$_4$H$_8$O)) amu. ^{1}H NMR (CDCl$_3$) δ: 3.8 (br), 3.3 (br), 1.8 (br) ppm. ^{13}C NMR (CDCl$_3$) δ: 153–144 (>30 peaks), 67.61 (m), 62.31 (m), 53.04 (m), 50.83 (m) ppm. UV–vis (ethanol), λ_{max}: 570 (sh), 390 (sh), 236 (sh), 217 nm.

Preparation of MgBr$_2$ · THF. In a flame-dried three-neck 500-mL round-bottom flask, under argon atmosphere, Mg (Alfa) (3 g, 0.12 mol) was suspended in dry THF (100 mL). The flask was cooled in an ice bath, and two equivalents of 1,2-dibromoethane (21.3 mL, 0.25 mol) was added to it. After a brief induction period, a vigorous reaction occurred. The reaction mixture was allowed to warm to room temperature and stirred for 48 h. The precipitated white solid was filtered using a Schlenk filter and washed with cold THF (50 mL). It was then recrystallized from hot THF (150 mL)–r.t.–ice-bath. The white crystals of MgBr$_2$ · THF were filtered under argon and dried under vacuum at r.t. for 16 h (wt., 15.94 g, 50% yield).

Reaction of C$_{60}$ with t-Butylmagnesium Chloride. In a flame-dried three-neck 25-mL round-bottom flask, under argon atmosphere, C$_{60}$ (30 mg, 4.16 × 10^{-5} mol) was suspended in dry THF (5 mL). Stirring was started, and 60 equivalents of commercial t-butylmagnesium chloride (2.0 M solution in Et$_2$O) (1.25 mL, 2.5 × 10^{-3} mol) was added all at once. Immediate reaction occurred, and a dark red-brown solution was obtained. The mixture was stirred at room temperature for 4 h. Then methyl iodide (1.5 mL) was added via a syringe. No immediate visible change occurred, and the mixture was allowed to stir overnight and was then quenched with water (5 mL). The mixture was then transferred to a centrifuge tube with the aid of more water (100 mL) and was shaken and centrifuged. The top clear-water–THF layer was decanted, and the brown-black residue in the tube was similarly washed two more times with water (100 mL each). Finally, the residue was dissolved in fresh THF and precipitated with water. After centrifugation, the residue was dried under vacuum at r.t. (wt., 60 mg).

IR (KBr) ν: 2930 (st), 2860 (sh, s), 1450 (m), 1390 (m), 1360 (m), 1310 (vw), 1230 (m), 1200 (m), 1155 (w) cm^{-1}. FABMS: clusters of peaks at 1440 (weak) ($C_{60}Bu_{10}Me_{10}$), 1410 (weak) ($C_{60}Bu_{10}Me_8$), 1368 ($C_{60}Bu_9Me_9$), four clusters consistent with the loss of four carbons, 1296 ($C_{60}Bu_8Me_8$), four clusters, 1224 ($C_{60}Bu_7Me_7$), four clusters, 1152 ($C_{60}Bu_6Me_6$), four clusters, 1080 ($C_{60}Bu_5Me_5$), four clusters, 1008 ($C_{60}Bu_4Me_4$), four clusters, 936 ($C_{60}Bu_3Me_3$), four clusters, 864 ($C_{60}Bu_2Me_2$), four clusters, 792 ($C_{60}BuMe$), four clusters, 720 (C_{60}) amu. ^{1}H NMR (THF-d_8) δ: 0.8–1.6 (v br) ppm, plus a few impurity peaks. TGA: wt. loss 50–450 °C, 50%; 50–580 °C, 64%. UV–vis (hexanes) λ_{max}: onset at 750; shoulders at 440, 415, 280, and 220; 205 nm.

Reaction of C_{60} with t-Butyllithium–Magnesium Bromide. In a flame-dried three-neck 25-mL round-bottom flask, under argon atmosphere, MgBr$_2 \cdot$ THF (0.64 g, 2.5 × 10^{-3} mol) was suspended in dry THF (3 mL). To this suspension, t-BuLi (0.7 M solution in pentane) (3.57 mL, 2.5 × 10^{-3} mol) was added via a syringe. The suspended salt dissolved immediately, and a clear solution was obtained. After stirring this solution at room temperature for 10 min, it was added, via a syringe, to C_{60} (30 mg, 4.16 × 10^{-5} mol) suspended in dry THF (2 mL) in a different flame-dried three-neck 25-mL round-bottom flask, also under argon atmosphere. The reaction mixture became dark green. After stirring at room temperature for 3 h, methyl iodide (1.5 mL) was added via a syringe. The mixture was stirred overnight and then quenched with water (5 mL). It was then transferred to a centrifuge tube with the aid of more water (100 mL). It was shaken and centrifuged. The top clear-water–THF layer was decanted, and the dark brown residue in the tube was transferred to a fritted filter funnel and washed repeatedly with water. Unlike the product from the t-butylmagnesium chloride reaction, the product was found to be much less soluble in THF. Finally, the residue was dried under vacuum at r.t. (wt., 110 mg).

IR (KBr) ν: 2930 (s), 2860 (sh, s), 1620 (vw), 1430 (s, br), 1370 (sh, s) cm^{-1}. FABMS: the fragmentation pattern is quite different from that of the product of the t-butylmagnesium chloride reaction; however, clusters of peaks present at 1440 (weak) ($C_{60}Bu_{10}Me_{10}$), 1410 (weak) ($C_{60}Bu_{10}Me_8$), 1368 (weak) ($C_{60}Bu_9Me_9$), 1296 (weak) ($C_{60}Bu_8Me_8$), 1224 (weak) ($C_{60}Bu_7Me_7$), 1152 (weak) ($C_{60}Bu_6Me_6$), several weak clusters, 792 ($C_{60}BuMe$), 720 (C_{60}) amu.

A small portion of the product (19 mg) was taken in a centrifuge tube and treated with THF (10 mL). The mixture was shaken on a mechanical shaker for 30 min and then centrifuged. The top brown solution was pipetted out and evaporated to yield a brown solid (wt., 3 mg). FABMS: same as previous. UV–vis (THF) λ_{max}: onset at 630, solvent cut off at 230 nm.

Reaction of C_{60} with t-Butyllithium. In a dry 25-mL two-neck flask, 60 equivalents of t-butyllithium in pentane was added under argon atmosphere to a suspension of 10 mg (1.4 × 10^{-5} mol) of C_{60} in 5 mL of THF with a syringe within 5 min. The formation of a dark brown suspension was observed. After 7 h of stirring, 1 mL of methyl iodide was added. After another 14 h, water was added to the brown suspension, and the precipitate was centrifuged and washed with water (twice). The brown solid was dissolved in THF (10 mL) and filtered through a frit. The THF was evaporated, and the product was dried under vacuum at r.t.

IR (KBr) ν: 2910 (m), 2860 (sh), 1450 (m, br), 1380 (sh, br), 1050 (s, br), 860 (vw), 420 (m, br) cm^{-1}. FABMS: clusters of peaks at 792 (C_{60}BuMe), 780 ($C_{60}Me_4$), 765 ($C_{60}Me_3$), 750 ($C_{60}Me_2$), 735 ($C_{60}Me$), 720 (C_{60}) amu. ^{1}H NMR (THF-d_8) δ: 1.4–2.7 (v br) ppm and signals of impurities.

Reaction of C_{60} with Phenyllithium–Magnesium Bromide. In a flame-dried three-neck 25-mL round-bottom flask, under argon atmosphere, $MgBr_2 \cdot$ THF (0.64 g, 2.5 × 10^{-3} mol) was suspended in dry THF (3 mL). To this suspension, PhLi (1.2 M solution in cyclohexane–ether) (2.08 mL, 2.5 × 10^{-3} mol) was added with a syringe. The suspended salt dissolved immediately, and a clear solution was obtained. After stirring this solution at room temperature for 35 min, it was added, with a syringe, to C_{60} (30 mg, 4.16 × 10^{-5} mol) suspended in dry THF (2 mL) in a different flame-dried three-neck 25-mL round-bottom flask, also under argon atmosphere. The reaction mixture became dark green. After stirring at room temperature for 4.5 h, methyl iodide (1.5 mL) was added with a syringe. The mixture was stirred overnight and then quenched with water (5 mL). It was then transferred to a centrifuge tube with the aid of more water (100 mL). It was shaken and centrifuged. The top yellow water–THF layer was decanted, and the dark red-brown oily residue in the tube was washed similarly once more with water. Finally, the residue (red-brown solid) was filtered and washed repeatedly with water and then dried under vacuum at r.t. (wt., 38 mg).

The ^{1}H NMR spectrum in CS_2–THF-d_8 indicated the presence of biphenyl in the product; there were multiplets at 7.2, 7.26, 7.36, and 7.5 ppm. The product was dissolved in CS_2 and precipitated with hexanes and centrifuged, and the supernatant was decanted. This process was repeated once more, and then finally the centrifuged precipitate was washed with fresh hexanes. The precipitate was shown (by MS) after drying to be mostly unreacted C_{60}. All of the organic rinsings were combined and evaporated to yield a brown solid, which was dried under vacuum (wt., 24 mg); redissolved in CS_2, and precipitated with ethanol. The precipitate was separated by centrifugation, and then the same process was repeated with fresh CS_2 and ethanol. This cleaning process was repeated one more time. Finally the precipitated brown material was rinsed with ethanol and then dried under vacuum at r.t.

^{1}H NMR (CS_2–THF-d_8) δ: 6.5–8.2 (v br, 5 H) (no biphenyl peaks present), 2.7–1.4 (v br, 3 H) ppm, plus a few impurity peaks. FABMS: clusters of peaks present at 1640 (weak) ($C_{60}Ph_{10}Me_{10}$), 1563 (weak) ($C_{60}Ph_9Me_{10}$), 1441 (weak) ($C_{60}Ph_8Me_7$), 1302 (weak) ($C_{60}Ph_6Me_8$), 1272 (weak) ($C_{60}Ph_6Me_6$), 1257 (weak) ($C_{60}Ph_6Me_5$), 1242 ($C_{60}Ph_6Me_4$), 1226, 1211, 1195

($C_{60}Ph_5Me_6$), 1180 ($C_{60}Ph_5Me_5$), 1165 ($C_{60}Ph_5Me_4$), 1150 ($C_{60}Ph_5Me_3$), 1120 ($C_{60}Ph_5Me$), 1118 ($C_{60}Ph_4Me_6$), 1103 ($C_{60}Ph_4Me_5$), 1088 ($C_{60}Ph_4Me_4$), 1073 ($C_{60}Ph_4Me_3$), 1058 ($C_{60}Ph_4Me_2$), 1043, 1026 ($C_{60}Ph_3Me_5$), 1011 ($C_{60}Ph_3Me_4$), 996 ($C_{60}Ph_3Me_3$), 981 ($C_{60}Ph_3Me_2$), 967, 951 ($C_{60}Ph_3$), 16 clusters of masses, 720 (C_{60}) amu. IR (KBr) ν: 3050, 3020 (both vw), 2950 (w, sh), 2900 (m), 2850 (m), 1600 (w), 1490 (w), 1445 (m), 1180 (vw), 1155 (vw), 730 (m), 695 (m) cm^{-1}. UV–vis (benzene) λ_{max}: onset at 620, solvent cut off at 280 nm.

Reaction of C_{60} with Phenylmagnesium Bromide. In a flame-dried three-neck 25-mL round-bottom flask, under argon atmosphere, C_{60} (30 mg, 4.16 × 10^{-5} mol) was suspended in dry THF (5 mL). Stirring was started, and 60 equivalents of commercial phenylmagnesium bromide (3.0 M solution in Et$_2$O) (0.83 mL, 2.5 × 10^{-3} mol) was added to it all at once. The mixture was stirred at room temperature for 4 h. Then methyl iodide (1.5 mL) was added via a syringe. The mixture was stirred overnight and then quenched with water (5 mL). It was then transferred to a centrifuge tube with the aid of more water (100 mL). It was shaken and centrifuged. The top clear-water–THF layer was decanted, and the brown-black residue in the tube was similarly washed two more times with water (100 mL each) and was dried under vacuum at r.t. (wt., 43 mg). It was then dissolved in CS$_2$ and transferred to a centrifuge tube (150 mL) and precipitated with ethanol. After centrifugation, the top reddish organic layer was decanted, and the brown precipitate was redissolved in CS$_2$ and reprecipitated with ethanol. After centrifugation, the brown precipitate was rinsed two times with fresh ethanol and then dried under vacuum at r.t.

FABMS: The fragmentation pattern looks quite different from the PhLi–MgBr$_2$ reaction case; however, there were clusters of peaks at 919 (weak) ($C_{60}Ph_2Me_3$), 889 (weak) ($C_{60}Ph_2Me$), 842 (weak) ($C_{60}PhMe_3$), 812 (weak) ($C_{60}PhMe$), 765 ($C_{60}Me_3$), 750 ($C_{60}Me_2$), 735 ($C_{60}Me$), 720 (C_{60}) amu. ^{1}H NMR (CS$_2$–THF-d_8) δ: 6.5–8.2 (v br), 2.7–1.4 (v br) ppm, plus a few impurity peaks. IR (KBr) ν: 3020 (vw), 2950 (m), 2910 (w, sh), 1620 (m, br), 1440 (m, br), 1180 (vw, sh), 670 (w), 565 (w), 525 (w), 430 (s) cm^{-1}.

Reaction of C_{60} with Phenyllithium. In a flame-dried three-neck 25-mL round-bottom flask, under argon atmosphere, C_{60} (30 mg, 4.16 × 10^{-5} mol) was suspended in dry THF (5 mL). Stirring was started, and 60 equivalents of commercial phenyllithium (1.2 M solution in cyclohexane–Et$_2$O) (2.08 mL, 2.5 × 10^{-3} mol) was added all at once. The mixture turned dark green immediately. It was stirred at room temperature for 4.5 h. Then methyl iodide (1.5 mL) was added with a syringe. The mixture was stirred overnight and then quenched with water (5 mL). It was then transferred to a centrifuge tube with the aid of more water (100 mL). It was shaken and centrifuged. The top clear-water–THF layer was decanted, and the brown-red residue in the tube was similarly washed two more times with water (100 mL each) and was dried under vacuum at r.t. It was then dissolved in CS$_2$ and transferred to a centri-fuge tube (150 mL) and precipitated with ethanol. After centrifugation, the supernatant was decanted, and the brown precipitate was redissolved in CS_2 and reprecipitated with ethanol. After centrifugation, the brown precipitate was rinsed two times with fresh ethanol and then dried under vacuum at r.t. (wt., 19 mg).

IR (KBr) ν: 3050, 3020 (vw), 2950 (w, sh), 2910 (m), 2850 (w, sh), 1600 (w), 1485 (w, sh), 1445 (m), 1370 (w), 1100 (m, sh), 735 (w, sh), 695 (m) cm^{-1}. FABMS: clusters at 1026 ($C_{60}Ph_3Me_5$), 1011 ($C_{60}Ph_3Me_4$), 996 ($C_{60}Ph_3Me_3$), 812 ($C_{60}PhMe$), 797 ($C_{60}Ph$), 735 ($C_{60}Me$), 720 (C_{60}) amu. ^{1}H NMR (CS_2-$CDCl_3$) δ: 6.5–8.3 (v br), 2.7–1.1 (v br) ppm, plus a few impurity peaks.

Reaction of C_{60} with $LiBH(Et)_3$. In a centrifuge vial that was sealed with a septum, 40 mg (5.6×10^{-5} mol) of C_{60} was dissolved in 10 mL of benzene under argon. To this solution, 0.06 mL of 1 M commercially available $LiBH(Et)_3$ "superhydride" in THF was added with a syringe. A dark precipitate was formed immediately. After 16 h of stirring, the dark solid was centrifuged and washed with benzene (3×5 mL) and THF (2×5 mL). The product was dried under vacuum for 1 week; yield 35 mg (87%, assuming the product is $LiC_{60}H$).

Anal.: Calcd. for $C_{60}HLi \cdot 6H_2O$: C, 86.22%; H, 1.57%; calcd for $C_{60}HLi \cdot 9H_2O$: C, 80.90%; H, 2.13%. Found: C, 80.97%; H, 1.58%. IR (KBr) ν: 3040 (w, sh), 3400 (br), 2960 (vw), 2925 (vw), 1485 (m, br), 1430 (m, br) 1125 (m), 1040 (sh), 1015 (sh), 860 (vw), 500 (m, br) cm^{-1}. FABMS: clusters of peaks 728 ($LiC_{60}H$), 727 (LiC_{60}), 720 (C_{60}) amu. ^{1}H NMR (pyridine-d_5): attempts to record a spectrum failed, possibly due to signal broadening and "signal dilution" (one atom out of a molecular weight of 728). ^{7}Li NMR (pyridine-d_5), standard 0.2 M LiCl in pyridine-d_5, r.t., δ: −0.186 ppm. ^{7}Li NMR–MAS (solid state), standard LiCl, r.t., δ: 1.250 ppm. ^{7}Li NMR (solid state) quantitative Li determination; 1.2 and 3.7 mg of LiCl as reference for 24 mg of compound: Calcd. for $C_{60}HLi$: 0.96% Li. Calcd. for $C_{60}HLi \cdot 9H_2O$: 0.79% Li. Found: 0.77 ± 0.16% Li. UV–vis (methanol) λ_{max}: 235 (sh), 205 nm.

Conclusions

Buckminsterfullerene, C_{60}, is an excellent electrophile. Our survey, so far, has shown that a variety of neutral and charged nucleophilic reagents, as well as dienes and dipoles, add smoothly to this spherical, electronegative carbon allotrope. In some cases we have been able to show via ESR spectroscopy that the addition is stepwise; the nucleophile first transfers an electron to C_{60}, and the resulting radical ionic salt collapses to product in a second step. The product of excess ethylenediamine addition is water soluble.

Acknowledgments

F. Wudl thanks the National Science Foundation for support through Grants DMR–88–20933 and CHE89–08323. H. Eckert thanks the NSF for Grants DMR 89–13738 and CHE 90–03542 (NMR instrumentation). A. Hirsch thanks the Deutsche Forschungsgemainschaft for a fellowship.

References

1. Krätschmer, W.; Lamb, L. D.; Fostiropoulos, K.; Huffman, D. R. *Nature (London)* **1990,** *347,* 354.

2. Taylor, R.; Hare, J. P.; Abdul-Sada, A. K.; Kroto, H. W. *J. Chem. Soc. Chem. Commun.* **1990,** 1423.

3. Bethune, D. S.; Meijer, G.; Tang, W. C.; Rosen, H. *J. Chem. Phys. Lett.* **1990,** *174,* 219.

4. Johnson, R. D.; Meijer, G.; Bethune, D. S. *J. Am. Chem. Soc.* **1990,** *112,* 8983.

5. Ajie, H.; Alvarez, M. M.; Anz, S. J.; Beck, R. D.; Diederich, F.; Fostiropoulos, K.; Huffman, D. R.; Krätschmer, W.; Rubin, Y.; Schriver, K. E.; Sensharma, D.; Whetten, R. L. *J. Phys. Chem.* **1990,** *94,* 8630.

6. Haufler, R. E.; Conceicao, J.; Chibante, L. P. F.; Chai, Y.; Byrne, N. E.; Flanagan, S.; Haley, M. M.; O'Brien, S. C.; Pan, C.; Xiao, Z.; Billups, W. E.; Ciufolini, M. A.; Hauge, R. H.; Margrave, J. L.; Wilson, L. J.; Curl, R. F.; Smalley, R. E. *J. Phys. Chem.* **1990,** *94,* 8634.

7. Lichtenberger, D. L.; Nebesny, K. W.; Ray, C. D.; Huffman, D. R.; Lamb, L. D. *Chem. Phys. Lett.* **1991,** *176,* 203.

8. Wragg, J. L.; Chamberlain, J. E.; White, H. W.; Krätschmer, W.; Huffman, D. R. *Nature (London)* **1990,** *348,* 623.

9. Wilson, R. J.; Meijer, G.; Bethune, D. S.; Johnson, R. D.; Chambliss, D. D.; de Vries, M. S.; Hunziker, H. E. *Nature (London)* **1990,** *348,* 621.

10. Yannoni, C. S.; Johnson, R. D.; Meijer, G.; Bethune, D. S.; Salem, J. R. *J. Phys. Chem.* **1991,** *95,* 9.

11. Frum, C. I.; Engleman, R. J.; Hedderich, H. G.; Bernath, P. F.; Lamb, L. D.; Huffman, D. R. *Chem. Phys. Lett.* **1991,** *176,* 504.

12. Allemand, P.-M.; Koch, A.; Wudl, F.; Rubin, Y.; Diederich, F.; Alvarez, M. M.; Anz, S. J.; Whetten, R. L. *J. Am. Chem. Soc.* **1991,** *113,* 1050.

13. Arbogast, J. W.; Darmanyan, A. P.; Foote, C. S.; Diederich, F. N.; Whetten, R. L.; Rubin, Y. *J. Phys. Chem.* **1991,** *95,* 11.

14. Tycko, R.; Haddon, R. C.; Dabbagh, G.; Glarum, S. H.; Douglass, D. C.; Mujsce, A. M. *J. Phys. Chem.* **1991,** *95,* 518.

15. Fleming, R. M.; Siegrist, T.; Marsh, P. M.; Hessen, B.; Kortan, A. R.; Murphy, D. W.; Haddon, R. C.; Tycko, R.; Dabbagh, G.; Mujsce, A. M.; Kaplan, M. L.; Zahurak, S. M. *Mater. Res. Soc. Proc.* **1991,** *206,* 352.

16. Haddon, R. C.; Schneemeyer, L. F.; Waszczak, J. V.; Tycko, R.; Dabbagh, G.; Kortan, A. R.; Muller, A. J.; Mujsce, A. M.; Rosseinsky, M.; Zahurak, S. M.; Thiel, F. A.; Raghavachari, K.; Elser, V. *Nature (London)* **1991,** *350,* 320.

17. Hawkins, J. M.; Lewis, T. A.; Loren, S. D.; Meyer, A.; Heath, J. R.; Shibato, Y.; Saykally, R. J. *J. Org. Chem.* **1990,** *55,* 6250.

18. Heymann, D. *Carbon* **1991,** *29,* 684.

19. Diederich, F.; Ettl, R.; Rubin, Y.; Whetten, R. L.; Beck, R. D.; Alvarez, M. M.; Anz, S. J.; Sensharma, D.; Wudl, F.; Khemani, K. C.; Koch, A. *Science (Washington, D.C.)* **1991,** *252,* 548.

20. Sunderlin, L. S.; Paulino, J. A.; Chow, J.; Kahr, B.; Ben-Amotz, D.; Squires, R. R. *J. Am. Chem. Soc.* **1991,** *113,* 5489.

21. Bausch, J. W.; Prakash, S. G. K.; Olah, G. A. *J. Am. Chem. Soc.* **1991,** *113,* 3205.

22. Hawkins, J. M.; Lewis, T. A.; Loren, S. D.; Meyer, A.; Heath, J. R.; Saykally, R. J.; Hollander, F. J. *Science (Washington, D.C.)* **1991,** *252,* 312.

23. Fagan, P. J.; Calabrese, J. C.; Malone, B. *Science (Washington, D.C.)* **1991,** *252,* 1160.

24. Koch, A.; Khemani, K. C.; Wudl, F. *J. Org. Chem.* **1991,** *56,* 4543.

Received August 28, 1991

Chapter 12

The Chemical Nature of C_{60} as Revealed by the Synthesis of Metal Complexes

Paul J. Fagan, Joseph C. Calabrese, and Brian Malone

Central Research and Development Department, E. I. du Pont de Nemours and Company, Inc., Experimental Station, P.O. Box 80328, Wilmington, DE 19880–0328

In this chapter we report our investigations of the chemical reactivity of C_{60} toward ruthenium and platinum reagents. The reagent $\{[\eta^5-C_5(CH_3)_5]-Ru(CH_3CN)_3\}^+(O_3SCF_3^-)$ reacts with C_{60} to form the complex $\{[\eta^5-C_5(CH_3)_5]Ru(CH_3CN)_2\}_x(C_{60})^{x+}(O_3SCF_3^-)_x$ ($x \approx 3$). The platinum reagent $[(C_6H_5)_3P]_2Pt(\eta^2-C_2H_4)$ reacts with C_{60} to form the complex $[(C_6H_5)_3P]_2Pt(\eta^2-C_{60})$, the structure of which has been determined by X-ray crystallography. This chemistry can be extended to prepare a new palladium compound and a platinum-coated C_{60} derivative, namely, $\{[(C_2H_5)_3P]_2Pt\}_6C_{60}$. We concluded from these reactions that the double bonds in C_{60} react like those in electron-poor alkenes and arenes rather than like those in benzene.

Very little is known about the chemical reactivity of the fullerenes (C_{60}, C_{70}, C_{84}, etc.) (*1–30*), which have only recently become available in large quantity (*1–11*). A pyridine-stabilized osmium tetroxide adduct of C_{60} was the first well-defined derivative known for these molecules (*25, 26*). Initially, it was thought by some that fullerenes were relatively inert to chemical attack and might behave more like benzene, which is chemically stabilized owing to its aromatic character (*21, 22, 31*). The true chemical nature of C_{60} (buckminsterfullerene) and the other fullerenes thus remained somewhat mysterious until we investigated the chemical reactivity of C_{60} toward ruthenium and platinum reagents (*24*).

Some evidence in the literature suggested that gas-phase metal complexes could be formed (*32–34*) and that C_{60} interacts with metal surfaces (*35–37*). We have been able to isolate a number of metal derivatives with the platinum complex $[(C_6H_5)_3P]_2Pt(\eta^2-C_{60})$ being characterized by a single-crystal X-ray analysis (*24*). Because the chemical properties of these platinum and

0097–6156/92/0481–0177$06.00/0

ruthenium reagents were known, we were able to conclude that C_{60} does not behave chemically like relatively electron-rich planar aromatic molecules such as benzene. Rather, the carbon–carbon double bonds of C_{60} react like those in very electron-deficient alkenes (or arenes) such as, for example, tetra-cyanoethylene. Thus, C_{60} should be capable of participating in reactions known for this class of compounds, including Diels–Alder reactions and nucleophilic substitution with electron-rich reagents. A similar perspective has since been offered by Diederich and Whetten (10), and Wudl and co-workers have evidence for these reactions (Wudl, F., University of California, Santa Barbara, personal communication). Indeed, the attack by electron-rich zero-valent platinum fragments is a type of nucleophilic substitution. This view-point suggests directions that should be taken in pursuing the chemistry of C_{60} and the other fullerenes.

Ruthenium Chemistry

The reaction of C_{60} with the organometallic reagent $\{[\eta^5\text{-}C_5(CH_3)_5]\text{-}Ru(CH_3CN)_3\}^+(O_3SCF_3{}^-)$ (38) was examined initially. When $\{[\eta^5\text{-}C_5(CH_3)_5]\text{-}Ru(CH_3CN)_3\}^+(O_3SCF_3{}^-)$ is reacted with relatively electron-rich planar arenes, the three coordinated acetonitrile ligands are displaced, resulting in strong, hexahapto-binding of ruthenium to the six-membered rings of the arene (Scheme I). In the presence of electron-poor alkenes, one of the acetonitrile ligands is displaced, and an olefin complex is formed. However, when given the choice between a phenyl ring or an alkene functionality such as in styrene, the ruthenium binds exclusively to the arene ring (38). If the six-membered rings of C_{60} behave like benzene, we would then expect all three acetonitrile ligands to be displaced from ruthenium, readily binding C_{60} in a hexahapto-fashion. Reactivity characteristic of an electron-poor olefin would be expected to dis-place just one acetonitrile ligand from ruthenium.

Reaction of C_{60} with a 10-fold excess of the reagent $\{[\eta^5\text{-}C_5(CH_3)_5]\text{-}Ru(CH_3CN)_3\}^+(O_3SCF_3{}^-)$ in CH_2Cl_2 at 25 °C over a period of 5 days yielded a brown precipitate (Reaction 1). The data we obtained for this compound suggested a formulation of $\{[\eta^5\text{-}C_5(CH_3)_5]Ru(CH_3CN)_2\}_x(C_{60})^{x+}(O_3SCF_3{}^-)_x$ with $x \sim 3$ (24). NMR spectroscopy demonstrated that two acetonitrile ligands were retained on each ruthenium, and this finding suggested that ruthenium was bound to an edge of the C_{60} cluster to fulfill its electron count (Figure 1). This reactivity was characteristic of an electron-poor alkene. We could obtain no further information on the structure of this complex. The spectra taken at room temperature were deceptively simple, and the ^{1}H NMR spectrum in CD_3NO_2 was composed of two singlets at δ 1.82 (broadened singlet, 15 H, η^5-$C_5(CH_3)_5$) and δ 2.62 (broadened singlet, 6 H, CH_3CN). The resonances were broadened, and the line widths decreased with increasing concentration of the compound. A process involving exchange of free and bound acetonitrile ligands might be occurring concurrent with migration of ruthenium on the C_{60} surface.

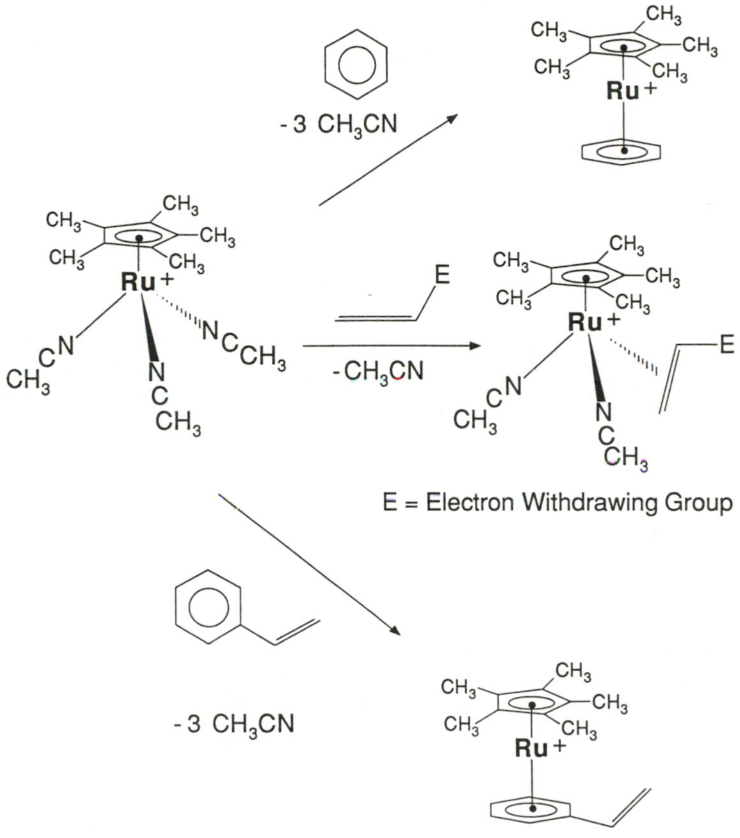

Scheme I. *Chemistry of $\{[\eta^5\text{-}C_5(CH_3)_5]Ru(CH_3CN)_3\}^+(O_3SCF_3^-)$.*

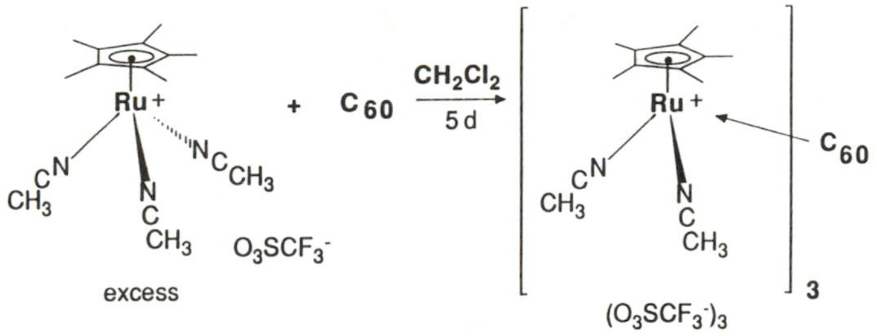

Reaction 1

Cooling the sample to $-35\ °C$ causes both the $\eta^5\text{-}C_5(CH_3)_5$ and acetonitrile resonances each to split into a forest of singlets (>15) (Fagan, P. J., unpublished results). The large number of resonances suggests the presence of several regioisomers, and restricted rotations about the Ru to C$_{60}$ alkene bonds would further increase the isomer count. Another possibility is that this compound may be a mixture of di-, tri-, and tetrasubstituted species, which exchange by dissociation of $[\eta^5\text{-}C_5(CH_3)_5]Ru(CH_3CN)_2^+$. Although this chemistry is complex, it did point to the fact that C$_{60}$ behaved chemically like an

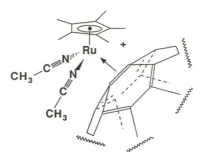

Figure 1. Proposed bonding for each $[\eta^5\text{-}C_5(CH_3)_5]Ru(CH_3CN)_2^+$ group in the complex $\{[\eta^5\text{-}C_5(CH_3)_5]Ru(CH_3CN)_2\}_x(C_{60})^{x+}(O_3SCF_3^-)_x$ (with x = 3). The double bond bound to Ru is expected to be parallel to the $C_5(CH_3)_5$ ligand, but is shown otherwise for clarity.

electron-poor alkene, and it led us to try reactions with platinum reagents, which are well-known to react with electron-poor alkenes and arenes to form dihapto-bound olefin complexes (39–45).

One final comment is in order regarding why C_{60} does not bind in a hexahapto-fashion and displace all three acetonitrile molecules from $[\eta^5\text{-}C_5(CH_3)_5]Ru(CH_3CN)_3^+$. Although C_{60} is suited for bonding in a dihapto-fashion (discussed later), it is not suited for hexahapto-bonding (and to some extent tetrahapto-bonding) because from above the "plane" of a six-membered ring in the molecule, the carbon p orbitals are tilted away from the center of the ring. Relative to a planar aromatic molecule, this tilt may weaken the overlap of the highest occupied and lowest unoccupied molecular orbitals of C_{60} with the ruthenium-centered unfilled and filled d-orbitals, respectively. In this case, acetonitrile is apparently a strong enough donor to prevent hexahapto-bonding. This is not to say that hexahapto-bonding is not possible with a weaker donor than acetonitrile, or by using another metal ligand fragment. Mass spectral evidence suggests that the species $(\eta^5\text{-}C_5H_5)Os(C_{60})^+$ exists in the gas phase (Shapley, J. R., University of Illinois, personal communication).

Platinum and Other Metals

Phosphine-stabilized zero-valent Ni, Pd, and Pt compounds not only react with electron-poor alkenes and arenes (39–45), but are also known to react with strained nonplanar alkenes (46). The double bonds in C_{60} belong to both of these classes. Coordination of metals to an alkene carbon–carbon double bond causes the four groups attached to it to splay back away from the metal (41). Thus, the native geometry of C_{60} is almost ideally constructed for dihapto-bonding to a transition metal (24). Reaction of $[(C_6H_5)_3P]_2Pt(\eta^2\text{-}C_2H_4)$ with a purple solution of C_{60} in toluene results in formation of an emerald green solution from which black microcrystals precipitated over the course of 2 h (Reaction 2). These could be isolated and recrystallized from tetrahydrofuran. The isolated yield of this compound was 85% based on the formulation $[(C_6H_5)_3P]_2Pt(\eta^2\text{-}C_{60})$ (tetrahydrofuran of crystallization was removed upon

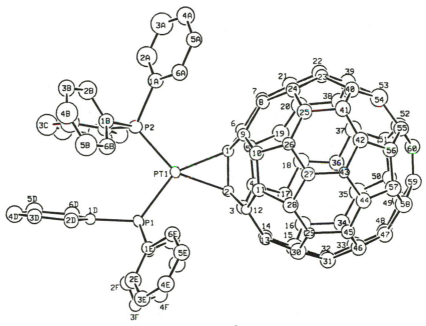

$$\text{Ph}_3\text{P} \diagdown \atop \text{Ph}_3\text{P} \diagup \text{Pt} \leftarrow \| \begin{array}{c} \text{CH}_2 \\ \text{CH}_2 \end{array} + \text{C}_{60} \xrightarrow[\text{toluene}]{-\text{CH}_2=\text{CH}_2} \begin{array}{c} \text{Ph}_3\text{P} \diagdown \\ \text{Ph}_3\text{P} \diagup \end{array} \text{Pt} \leftarrow \text{C}_{60}$$

Reaction 2

Figure 2. Drawing of $[(C_6H_5)_3P]_2Pt(\eta^2\text{-}C_{60})$. Hydrogen atoms on phenyl groups are omitted. (Reproduced with permission from reference 24. Copyright 1991 American Association for the Advancement of Science.)

drying in vacuo). The elemental analysis and the spectroscopic properties of this complex were consistent with this formulation (*24*).

The compound was slightly soluble in tetrahydrofuran, giving a dark green solution, and a single crystal of the complex $[(C_6H_5)_3P]_2Pt(\eta^2\text{-}C_{60})$ · C_4H_8O suitable for an X-ray analysis was grown from this solvent by slow evaporation. Although only a limited amount of data could be obtained, the structure was solved (*24*) and is shown in Figure 2. The structure confirmed that the platinum atom was bound in a dihapto-manner to one of the C_{60} carbon–carbon double bonds. The structure of the C_{60} cage is similar to that reported by Hawkins et al. (*25*) for the osmium tetroxide derivative (*t*-BuC$_5$H$_4$N)$_2$OsO$_4$C$_{60}$. Accuracy in both the osmium tetroxide and platinum structures was rather poor owing to the limited amount of data available, and the atoms of the carbon cages could only be refined isotropically (*24, 25*). The experimental error of the C–C bond lengths in both structures was on the order of ±0.03 to ±0.04 Å.

The two types of bonds within C_{60} are those exo to the five-membered rings (at the fusions of two six-membered rings) and those within the five-membered rings (at the junctions of five- and six-membered rings). Theoretical calculations suggest that the bonds exo to the C_5 rings should have the most double-bond character and be shorter than bonds within the C_5 ring (12). In the platinum derivative, if the C—C bonds involved with Pt binding are excluded, the average value for the bonds exo to the five-membered rings is 1.389 ($\pm$0.043) Å (range 1.318 (29) to 1.471 (35) Å), whereas the bonds within the five-membered rings average to 1.445 ($\pm$0.040) Å (range 1.365 (30) to 1.534 (39) Å). (The errors for bond lengths are the estimated standard deviations for the experimental bond-length determination; the error quoted for the average values is the standard deviation of the average.) These averages compare well with those found in $(t\text{-}BuC_5H_4N)_2OsO_4C_{60}$, namely, 1.388 (9) and 1.432 (5) Å, respectively. Considering the accuracy of the osmium structure, we believe the errors reported by these authors for these averages are not meaningful (25). In our original paper (24), we estimated an error in these averages that was the approximate error in the C—C bond-length determinations ($\pm$0.03 Å). We now think the best and most conservative error chosen should be the standard deviation of the calculated average of the bond measurements and have revised the errors for the averages in the platinum structure to approximately $\pm$0.04 Å.

We cannot know if the range of bond values found for each of the two classes of bonds in C_{60} is owing to the experimental inaccuracy of the structural determination, or if in fact there is a real range of values for these bonds in the solid state. Platinum is bound to the C_{60}, and there are intermolecular C_{60} to C_{60} contacts (which may cause some electronic reorganization owing to charge transfer near such contacts); consequently, one "true value" or even a small range of values may not exist for each of the two classes of bonds in the structure. The error reported for the osmium tetroxide derivative was the standard error (standard deviation divided by the square root of the number of measurements) (Hawkins, J., University of California, Berkeley, personal communication). We suggest that when it is not known or it cannot be determined experimentally if a single "true value" or at least a very small range of values exists for a set of bond measurements, calculation of a standard error is not valid, and the standard deviation is a more meaningful number.

As can be seen in Figure 2, the bis(triphenylphosphine)platinum group coordinates at the fusion of two six-membered rings, consistent with these bonds being shorter and having the most double-bond character. This is the same site of reactivity seen for the osmium tetroxide derivative (25). Coordination of platinum to C_{60} pulls out the two attached carbons from the C_{60} framework. All of the other carbons have an average distance to the C_{60} centroid of 3.53 Å, but the two attached to Pt are 3.63 and 3.73 Å from the centroid. This result is also reflected in the planarity of the rings, which are all planar to within 0.03 to 0.05 Å, except for those rings associated with platinum binding. The coordination sphere about Pt is very similar to that seen in related alkene complexes (24). For example, in Figure 3, the coordination sphere about Pt in $[(C_6H_5)_3P]_2Pt(\eta^2\text{-}C_{60})$ is compared to that for $[(C_6H_5)_3P]_2Pt(\eta^2\text{-}C_2H_4)$ (45). The triphenylphosphine ligands in the C_{60} complex are bent back somewhat more (P—Pt—P = 102.4 (2)° versus 111.60 (7)° for the ethylene complex), a condition that can be attributed in part to the slightly greater steric bulk of C_{60} relative to ethylene.

We have just begun to investigate the synthesis of other metal derivatives of C_{60}. A number of related compounds have now been prepared, including the complexes $[(C_2H_5)_3P]_2Pt(\eta^2\text{-}C_{60})$ and $[(C_6H_5)_3P]_2Pd(\eta^2\text{-}C_{60})$ (Fagan, P.

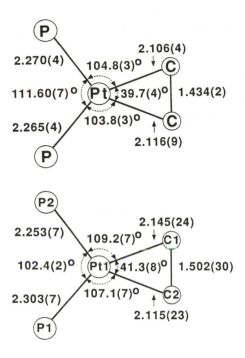

Figure 3. Comparison of the platinum coordination spheres in [(C$_6$H$_5$)$_3$-P]$_2$Pt(η^2-C$_{60}$) (bottom) (24) and [(C$_6$H$_5$)$_3$P]$_2$Pt(η^2-C$_2$H$_4$) (top) (45). (Reproduced with permission from reference 24. Copyright 1991 American Association for the Advancement of Science.)

J., unpublished results). We have also prepared a platinum-coated C$_{60}$ derivative, namely, {[(C$_2$H$_5$)$_3$P]$_2$Pt}$_6$C$_{60}$ (47). This metal-coated derivative contains six platinum atoms in an octahedral array about the C$_{60}$ framework (as determined by spectroscopy and a single-crystal X-ray analysis). All the platinum atoms are bound in a dihapto-fashion at the fusions of two six-membered rings, and excluding the ethyl groups, this molecule is a rare example of a compound with T$_h$ point-group symmetry. Other multiply-substituted derivatives of C$_{60}$ have been reported (23–30), but they usually occur as mixtures and have not (as of September 1991) been structurally characterized. This result may help in answering fundamental questions concerning C$_{60}$, such as what, if any, geometrical preferences or electronic directing effects guide the substitution chemistry. It also shows that with proper control of electronic and steric factors, the synthesis of a single well-defined multiply-substituted isomer of C$_{60}$ can be accomplished in good yield.

Conclusion

We have been able to prepare some of the first metal derivatives of C$_{60}$, and these are also some of the best characterized derivatives that are known.

Clearly, the metal chemistry of C_{60} and the other fullerenes will be quite extensive. Not only will the inorganic solution chemistry be vast, but a host of both binary and ternary solid-state inorganic compounds are likely to be synthesized. There is the potential that metal compounds of C_{60} will make important contributions to catalytic and material sciences. Perhaps the most important observation from the organometallic chemistry is that the double bonds of C_{60} behave like those in electron-poor arenes and alkenes. This observation points the way for not only metal chemistry, but organic reactions as well.

Acknowledgments

We thank Edd Holler for chromatographic purification of C_{60}. We thank Will Marshall and Ron Davis for technical assistance.

References

1. Krätschmer, W.; Lamb, L. D.; Fostiropoulos, K.; Huffman, D. R. *Nature (London)* **1990**, *347*, 354.

2. Taylor, R.; Hare, J. P.; Abdul-Sada, A. K.; Kroto, H. W. *J. Chem. Soc. Chem. Commun.* **1990**, 1423.

3. Kroto, H. W.; Heath, J. R.; O'Brien, S. C.; Curl, R. F.; Smalley, R. E. *Nature (London)* **1985**, *318*, 162.

4. Stoddart, F. J. *Angew. Chem.* **1991**, *103*, 71.

5. Kroto, H. *Pure Appl. Chem.* **1990**, *62*, 407.

6. Kroto, H. *Science (Washington, D.C.)* **1988**, *242*, 1139.

7. Krätschmer, W.; Fostiropoulos, K.; Huffman, D. R. *Chem. Phys. Lett.* **1990**, *170*, 167.

8. Hare, J. P; Kroto, H. W.; Taylor, R. *Chem. Phys. Lett.* **1991**, *177*, 394.

9. Cox, D. M.; Behal, S.; Disko, M.; Gorun, S. M.; Greaney, M.; Hsu, C. S.; Kollin, E. B.; Millar, J.; Robbins, J.; Sherwood, R. D.; Tindall, P. *J. Am. Chem. Soc.* **1991**, *113*, 2940.

10. Diederich, F.; Whetten, R. L. *Angew. Chem. Int. Ed. Engl.* **1991**, *30*, 678.

11. Miller, J. S. *Adv. Mater.* **1991**, *3*, 262.

12. Scuseria, G. E. *Chem. Phys. Lett.* **1991**, *176*, 423, and references therein.

13. Haddon, R. C.; Schneemeyer, L. F.; Waszczak, J. V.; Glarum, S. H.; Tycko, R.; Dabbagh, G.; Kortan, A. R.; Muller, A. J.; Mujsce, A. M.; Rosseinsky, M. J.; Zahurak, S. M.; Makhija, A. V.; Thiel, F. A.; Raghavachari, K.; Cockayne, E.; Elser, V. *Nature (London)* **1991**, *350*, 46.

14. Yannoni, C. S.; Johnson, R. D.; Meijer, G.; Bethune, D. S.; Salem, J. R. *J. Phys. Chem.* **1991**, *95*, 9.

15. Allemand, P.-M.; Koch, A.; Wudl, F.; Rubin, Y.; Diederich, F.; Alvarez, M. M.; Anz, S. J.; Whetten, R. L. *J. Am. Chem. Soc.* **1991**, *113*, 1050.

16. Wasielewski, M. R.; O'Neil, M. P.; Lykke, K. R.; Pellin, M. J.; Gruen, D. M. *J. Am. Chem. Soc.* **1991**, *113*, 2774.

17. Allemand, P.-M.; Srdanov, G.; Koch, A.; Khemani, K.; Wudl, F.; Rubin, Y.; Diederich, F.; Alvarez, M. M.; Anz, S. J.; Whetten, R. L. *J. Am. Chem. Soc.* **1991**, *113*, 2780.

18. Yannoni, C. S.; Bernier, P. P.; Bethune, D. S.; Meijer, G.; Salem, J. R. *J. Am. Chem. Soc.* **1991**, *113*, 3190.

19. Lichtenberger, D. L.; Nebesny, K. W.; Ray, C. D.; Huffman, D. R.; Lamb, L. D. *Chem. Phys. Lett.* **1991**, *176*, 203.

20. Tycko, R.; Haddon, R. C.; Dabbagh, G.; Glarum, S. H.; Douglass, D. C.; Mujsce, A. M. *J. Phys. Chem.* **1991**, *95*, 518.

21. Fowler, P. *Nature (London)* **1991**, *20*, 350.

22. Kroto, H. W. *Nature (London)* **1987**, *329*, 529.

23. Selig, H.; Lifschitz, C.; Peres, T.; Fischer, J. E.; McGhie, A. R.; Romanow, W. J.; McCauley, J. P., Jr.; Smith, A. B. III, *J. Am. Chem. Soc.* **1991**, *113*, 5475.

24. Fagan, P. J.; Calabrese, J. C.; Malone, B. *Science (Washington, D.C.)* **1991**, *252*, 1160.

25. Hawkins, J. M.; Meyer, A.; Lewis, T. A.; Loren, S.; Hollander, F. J. *Science (Washington, D.C.)* **1991**, *252*, 312.

26. Hawkins, J. M.; Lewis, T. A.; Loren, S. D.; Meyer, A.; Heath, J. R.; Shibato, Y.; Saykally, R. J. *J. Org. Chem.* **1990**, *55*, 6250.

27. Haufler, R. E.; Conceicao, J. ; Chibante, L. P. F. ; Chai, Y.; Byrne, N. E.; Flanagan, S.; Haley, M. M.; O'Brian, S. C.; Pan, C.; Xiao, Z.; Billups, W. E.; Ciufolini, M. A.; Hauge, R. H.; Margrave, J. L.; Wilson, L. J.; Curl, R. F.; Smalley, R. E. *J. Phys. Chem.* **1990**, *94*, 8634.

28. Holloway, J. H.; Hope, E. G.; Taylor, R.; Langley, G. J.; Avent, A. G.; Dennis, T. J.; Hare, J. P.; Kroto, H. W.; Walton, D. R. M. *J. Chem. Soc. Chem. Commun.* **1991**, 966.

29. Bausch, J. W.; Prakash, G. K. S.; Olah, G. A.; Tse, D. S.; Lorents, D. C.; Bae, Y. K.; Malhotra, R. *J. Am. Chem. Soc.* **1991**, *113*, 3205.

30. Krusic, P. J.; Wasserman, E.; Parkinson, B. A.; Malone, B.; Holler, E. R., Jr. *J. Am. Chem. Soc.* **1991**, *113*, 6274.

31. Amic, D.; Trinajstic, N. *J. Chem. Soc. Perkin Trans. 2* **1990**, 1595.

32. Heath, J. R.; O'Brien, S. C.; Zhang, Q.; Liu, Y; Curl, R. F.; Kroto, H. W.; Tittel, F. K.; Smalley, R. E. *J. Am. Chem. Soc.* **1985,** *107,* 7779.

33. Weiss, F. D.; Elkind, J. L.; O'Brien, S. C.; Curl, R. F.; Smalley, R. E. *J. Am. Chem. Soc.* **1988,** *110,* 4464.

34. Roth, L. M.; Huang, Y.; Schwedler, J. T.; Cassady, C. J.; Ben-Amotz, D. ; Kahr, B.; Freiser, B. S. *J. Am. Chem. Soc.* **1991,** *113,* 6298.

35. Garrell, R. L.; Herne, T. M.; Szafranski, C. A.; Diederich, F.; Ettl, F.; Whetten, R. L. *J. Am. Chem. Soc.* **1991,** *113,* 6302.

36. Wilson, R. J.; Meijer, G.; Bethune, D. S.; Johnson, R. D.; Chambliss, D. D.; de Vries, M. S.; Hunziker, H. E.; Wendt, H. R. *Nature (London)* **1990,** *348,* 621.

37. Wragg, J. L.; Chamberlain, J. E.; White, H. W.; Krätschmer, W.; Huffman, D. R. *Nature (London)* **1991,** *348,* 623.

38. Fagan, P. J.; Ward, M. D.; Calabrese, J. C. *J. Am. Chem. Soc.* **1989,** *111,* 1698.

39. Cook, C. D.; Jauhal, G. S. *J. Am. Chem. Soc.* **1968,** *90,* 1464.

40. Browning, J.; Green, M.; Penfold, B. R.; Spencer, J. L.; Stone, F. G. A. *J. Chem. Soc. Chem. Commun.* **1973,** 35.

41. Ittel, S. D.; Ibers, J. A. *Adv. Organomet. Chem.* **1976,** *14,* 33.

42. Osborne, R. B.; Ibers, J. A. *J. Organomet. Chem.* **1982,** *232,* 371.

43. Francis, J. N.; McAdam, A.; Ibers, J. A. *J. Organomet. Chem.* **1971,** *29,* 149.

44. Bombieri, G.; Forsellini, E.; Panattoni, C.; Graziani, R.; Bandolini, G. *J. Chem. Soc. A* **1970,** 1313.

45. Cheng, P. T.; Nyburg, S. C. *Can. J. Chem.* **1972,** *50,* 912.

46. Morokuma, K.; Borden, W. T. *J. Am. Chem. Soc.* **1991,** *113,* 1912, and references therein.

47. Fagan, P. J.; Calabrese, J. C.; Malone, B. *J. Am. Chem. Soc.* **1991,** in press.

Received August 27, 1991

Author Index

Allemand, P.-M., 161
Anderson, William, 117
Bae, Young K., 127
Becker, Christopher H., 127
Bethune, Donald S., 107
Briant, Clive E., 41
Calabrese, Joseph C., 177
Cox, David E., 55
Cox, Donald M., 117
Creegan, Kathleen M., 117
Dabbagh, G., 25
Day, Cynthia S., 41
Day, Victor W., 41
Duclos, S. J., 71
Eckert, H., 161
Fagan, Paul J., 177
Fischer, John E., 55
Fleming, R. M., 25,71
Glarum, S. H., 71
Gorun, Sergiu M., 41
Greaney, Mark A., 41
Haddon, R. C., 25,71
Hammond, George S., vii, ix
Hawkins, Joel M., 91
Heath, James R., 1
Hebard, A. F., 71
Heiney, Paul A., 55
Hessen, B., 25
Hirsch, A., 161
Johnson, Robert D., 107
Jusinski, Leonard E., 127
Khemani, K. C., 161

Koch, A., 161
Kortan, A. R., 25
Kuck, Valerie J., vii
Lewis, Timothy A., 91
Loren, Stefan, 91
Lorents, Donald C., 127
Luzzi, David E., 55
Malhotra, Ripudaman, 127
Malone, Brian, 177
Marsh, P., 25
Martella, David J., 117
Meijer, Gerard, 107
Meyer, Axel, 91
Murphy, D. W., 71
Palstra, T. T. M., 71
Ramirez, A. P., 71
Rosseinsky, M. J., 71
Salem, Jesse R., 107
Sherwood, Rexford D., 117
Siegrist, T., 25,71
Smalley, R. E., 141
Srdanov, G., 161
Suzuki, T., 161
Tindall, Paul, 117
Tse, Doris S., 127
Tycko, R., 25,71
Upton, Roger M., 41
Wachsman, Eric D., 127
Webb, H. M., 161
Wudl, F., 161
Yannoni, Costantino S., 107

Affiliation Index

AT&T Bell Laboratories, 1,71
Brookhaven National Laboratory, 41
Chemical Design Ltd., 25
Crystalytics Company, 25
E. I. du Pont de Nemours and Company, Inc., 177
Exxon Chemicals Co., 117
Exxon Research and Engineering Company, 25,117

IBM Almaden Research Center, 107
Lehigh University, 117
Rice University, 141
SRI International, 127
University of California— Berkeley, 1,91
University of California— Santa Barbara, 161
University of Pennsylvania, 41

Subject Index

A

Absorption measurement, antisymmetric
  stretch and associated bending hot
  band of C_7 cluster, 14,15f
Addition reactions, C_{60}, 162
Adducts of C_{60}, formed with primary and
  secondary amines, 167
Alkali-intercalated C_{60}, preparation and
  structure, 64
Alkali fullerides, structural studies, 80
Alkali metal doped C_{60}
  band structure calculations, 82
  conductivity, 76
  structural studies, 80
  superconductivity, 77–79
Alkali metal doping, C_{60} and C_{70},
  apparatus and procedure, 76
Amine addition, C_{60}, 162–166
Ammonia titration, boronated
  fullerenes, 147
Anthracene, osmylation, 92,94f
Antisymmetric stretch
  C_7 cluster, 14
  linear C_9, 12f
  triplet linear C_4, 11f

B

Band structure calculations, alkali metal
  doped C_{60}, 82
Bending modes
  effect on isomerization pathways, 13
  linear carbon clusters, 10
Bending potential, C_3, 14
Binomial distribution, fullerene
  isotopomers, 2–3
Bond angles, osmylated C_{60}, 99,102f
Bond lengths
  C_{60} compounds, 48
  C_{60}–platinum complex, 181
  osmylated C_{60}, 100,102,103f
Bonding
  C_{60}, 97
  C_{60}–ruthenium complex, 180
  intermolecular, solid C_{60}, 63
  linear carbon clusters, 10

Boronated fullerenes
  ammonia titration, 147
  laser-shrinking, 149
tert -Butylamine, addition to C_{60}, 165

C

^{13}C NMR spectroscopy
  C_{60}–C_{70}, 108
  use in determining mechanism of
    formation of fullerenes, 110
C–C bond length refinement, C_{60}, 47
C_{60}
  2 + 4 cycloaddition, 166
  addition reactions, 162–166
  adducts formed with primary and
    secondary amines, 167
  amine addition, 162–166
  analysis of structure, 99
  and C_{60} monoanion radical, IR spectrum, 44
  anion radical, ESR spectrum, 42
  as a plastic crystal, 55
  bond lengths, 48,97
  bond-length refinement, 47
  brief history of discovery, 25
  ^{13}C NMR spectrum, liquid state, 108–111
  charged nucleophile addition, 165–166
  chemical reactivity, survey, 161–173
  crystal-packing environment, 47
  crystalline close-packed plane from
    idealized $Fm\overline{3}$ structure, 84
  crystalline maximal subgroups,
    icosahedral point group $m\,\overline{3}5$, 26,27f
  dienophile reactivity of double bonds, 165
  dipolar cycloaddition, 165–166
  dipolarophile nature, 165
  doped orientation of C_{60} molecule, 30
  doped preparation and structure, 64
  doped films, microwave loss as function
    of temperature, 79
  doped with alkali metals, 76–82
  doped with Cs, X-ray powder
    diffractogram, 64,65f
  doped with potassium, 76–81
  doped with rubidium, 77–81
  doping, possible ways, 142
  electrocyclic reactions, 162–163

C$_{60}$—*Continued*
ESR spectrum of anion radical, 42
field-ionization mass spectrum, 129
formation within carbon arc, 1
fullerite, thermogravimetric analysis, 135
Grignard addition, 162–163
idealized symmetry model, 27–28
IR spectrum, 44
isotopically resolved mass spectra, 3
mass chromatogram of osmate esters,
    93,95*f*
molecule crystal structure and symmetry, 26
molecule interstitial sites, 73*f*
molecule motion, 30
molecule packing models, 92,93*f*
negative ion chemical ionization mass
    spectrum, 162,164*f*
nucleophilic addition, 162
one- and two-dimensional NMR studies,
    107–113
organolithium reagent addition, 162–163
orbital energies used to calculate
    electronic transitions, 43
orientational order, 56
osmate esters, mass chromatogram,
    93,95*f*
osmium tetroxide adduct, ORTEP drawing,
    97,98*f*,99*f*
osmylated, *See* Osmylated C$_{60}$
osmylation, 92,94*f*
pentane solvates, 31
plasma desorption mass spectrometry, 118
platinum complex, bond lengths, 181
production of gram quantities, 128
production of macroscopic amounts, 153
pure, electrical properties, 142
purification, 167
radical anions, HOMO–LUMO
    orbitals, 45
reaction with metals, 177
reaction with propylamine, ESR spectra,
    162,164*f*
reaction with Pt complexes, 180
reaction with Ru complexes, 178
ruthenium complex, bonding, 180
small molecule synthesis supported by
    isotopic scrambling and NMR
    experiments, 4
soccer-ball geometry supported by ^{13}C
    NMR spectrum, 110

C$_{60}$—*Continued*
solid
    compressibility, 63
    electronic and molecular structure, 82
    native defect, 57
solid-state NMR spectra, 111–113
solvated, single-crystal X-ray
    structure, 41–51
solvates, pentane, 31
structure analysis, 99
sublimed crystals, 29
sublimed films, 123
temperature-programmed oxidation, 136–137
thermal evolution profiles, 131,134*f*
thin films, preparation and analysis, 73
C$_{70}$
^{13}C NMR spectrum, liquid state, 108–111
field-ionization mass spectrum, 129
one- and two-dimensional NMR studies,
    107–113
pentane solvates, 31
production of gram quantities, 128
production of macroscopic amounts, 153
radical anions, HOMO–LUMO orbitals, 45
solvates, 31
structure elucidated by ^{13}C NMR
    spectrum, 110
sublimation, 31
sublimed crystals, 29
sublimed films, 123
thermal evolution profiles, 131,134*f*
thin films, preparation and analysis, 73
Carbon-arc synthesis of fullerenes
    determination of initial reactants, 2
    significance, 1
    See also Condensation of carbon
Carbon atom, initial reactant for
    production of fullerenes, 4
Carbon clusters
    FTICR apparatus for studying, 145
    involved in carbon-arc fullerene synthesis, 4
    molecular parameters, 19
    See also Linear carbon clusters
Carbon condensation, equilibrium constant
    of isomerization reaction, 11
Carbon species, ejected from graphite rod
    by carbon arc, 2
Cesium-doped C$_{60}$
    cube face, 65,66*f*
    X-ray powder diffractogram, 64,65*f*

Charged nucleophile addition, C_{60}, 165–166

Chemical model, fullerene synthesis, 4

Chemical reactivity
C_{60}, survey, 161–173
fullerenes with metal complexes, 177

Chromatographic analysis, crude reaction mixture from osmylation of C_{60}, 96

Close contact, osmylated C_{60} molecules in solid state, 97,101f

Closed-shell synthesis model, fullerenes, 17

Cluster growth in one dimension, carbon condensation, 5–6

Clusters, *See* Carbon clusters

Collision complexes
calculation of lifetimes, 7–9
physics, 5

Compressibility, solid C_{60}, 63

Computer modeling, structure of solvated pure C_{60}, 46

Condensation of carbon
equilibrium constant of isomerization reaction, 11
gas, mechanism of formation of fullerenes, 110
monocyclic rings formed, 10
three-dimensional clusters formed, 16
vapor, various steps, 5–6
See also Carbon-arc synthesis

Conductivity
alkali metal doped C_{60}, 71–86
apparatus for measuring, 74
thin films of C_{60} and C_{70}, 73

Conductors, effect of pi orbitals, 71

Crystal-packing environment, C_{60}, 47

Crystal structure
C_{60}, 26,58
osmylated C_{60}, 97

Crystal(s)
10-sided, formed by twinning, C_{60} and C_{70} pentane solvates, 33
C_{60} and C_{70}, produced by pentane solvation, 31
grown from hexanes, packing models, 92
sublimed, C_{60} and C_{70}, 29

Crystalline C_{60}
close-packed plane from idealized $Fm\bar{3}$ structure, 84
maximal subgroups, icosahedral point group $m\,\bar{3}5$, 26,27f

Crystalline fullerenes, 25–37

Cube face, Cs-doped C_{60}, 65,66f

Cubic data-set refinement, solvated C_{60} structure, 46

Cycloaddition, 2 + 4, C_{60}, 165–166

Cyclohexane, fit in C_{60} octahedral and tetrahedral voids, 47

D

Defect in solid C_{60}, 57

Degenerate bending modes, *See* Bending modes

Desorption
bulk fullerene extract, 131
fullerenes from raw soot, 130

Dienophile reactivity, C_{60} double bonds, 165

Differential thermal analysis, fullerenes, 121

Diffraction, pure C_{60}, 29

Diffraction patterns, indexed on basis of unit cell, 31,32f

Dipolar cycloaddition, C_{60}, 165–166

Dipolarophile nature, C_{60}, 165

Dispersions, vacant bands, C_{60}, 84,85f

Doped C_{60}, *See* C_{60}, doped

Doped fullerenes
nomenclature and symbolism, 142
reactivity, 155

Doping
C_{60}, possible ways, 142
C_{60} and C_{70} with alkali metals, 76
effect on electrical properties, 141
fullerene cage, 143
fullerenes, three methods, 141–157
inside fullerene cage, 151
preformed fullerenes, 156
See also Intercalation

E

Electrocyclic reactions, C_{60}, 162–163

Electron diffraction patterns, C_{60} crystals, 58,59f

Electron spin resonance spectroscopy
C_{60} anion radical, 42
naphthalene–naphthalenide anion, 42
reaction of C_{60} with propylamine, 162,164f

Electronic properties, effect of directionality of pi orbitals, 71

Electronic structure
  C_{60} and C_{70} anions, 42
  solid C_{60}, EHT band structure
    calculations, 82
Electronic transitions of C_{60}, C_{60} orbital
  energies used to calculate, 43
Energies, vacant bands arising from fcc
  lattice of C_{60}, 83
Energy level expressions, for
  determination of carbon-cluster
  molecular parameters, 19
Equilibrium constant, isomerization
  reaction, carbon condensation, 11
Equilibrium ratios, dependence on
  temperature, chains and monocyclic
  rings, 12,13f
Ethylenediamine, addition to C_{60}, 162
Extraction, fullerenes from raw soot, 120

F

Field-ionization mass spectrometry,
  fullerenes, 128–129
Fourier transform ion cyclotron resonance
  (FTICR) apparatus, details, 145
Fourier transform ion cyclotron resonance
  (FTICR) mass spectrum, laser-vaporized
  target rods, 154
Fragmentation of spheroidal carbon shells,
  mechanisms, 17
Fullerene cage
  doping inside, 151
  doping outside, 143
Fullerene ions, electronic structure and
  production of bulk quantities, 42
Fullerene synthesis, chemical model, 4
Fullerenes
  boronated
    ammonia titration, 147
    laser-shrinking, 149
  brief history of discovery, 25
  chemical reactivity, 161
  closed-shell synthesis model, 17
  crystalline, 25–37
  differential thermal analysis, 121
  doped, nomenclature and symbolism, 142
  doping, three methods, 141–157
  extracted from raw soot, 120
  formation within carbon arc, 1,16
  mass spectrometry, 128

Fullerenes—Continued
  mechanism of formation, 110
  metals trapped inside cage, 152
  open-shell synthesis model, 16
  plasma desorption mass spectrometry, 118
  preformed, doping, 156
  production of gram quantities, 128
  proposed structures, 107
  SALI spectra, 130,132f
  separation by extraction, 120
  separation by sublimation, 122
  temperature-programmed oxidation,
    136–137
  thermal desorption, 130
  thermal properties, 120
  thermogravimetric analysis, 120,135
Fullerite (solid fullerenes), See Fullerenes

G

Graphite targets, laser vaporization, 153
Grignard addition, C_{60}, 162–163
Growth mechanism, spheroidal carbon
  shells, 18

H

Harmonic vibrational frequencies, linear
  carbon clusters, 9t
Helium-doped fullerenes, production, 156
Heteroatom as part of fullerene, cage, 144
High-resolution powder diffraction profiles
  C_{60} at 300 and 11K, 61,62f
  solid C_{60}, 57
Highest occupied molecular orbitals,
  symmetry, small carbon clusters, 13,14f
HOMO–LUMO orbitals, C_{60} and C_{70} radical
  anions, 45
HPLC trace, crude reaction mixture from
  osmylation of C_{60}, 96

I

Icosahedral point group $m\,\overline{3}5$, crystalline,
  maximal subgroups, 26,27f
Insulators, effect of doping, 141
Intercalated C_{60}, See C_{60}, doped
Intercalation
  procedure, C_{60}, 64
  See also Doping

Intermediates involved in carbon-arc
   fullerene synthesis, 4
Intermolecular bonding, solid C_{60}, 63
Internal energy of energized molecule,
   collision complex, 5
Interstitial sites, C_{60} molecules, 73f
IR spectrum
   C_{60} and C_{60} monoanion radical, 44
   C^7 cluster, 14,15f
   singlet linear C_9, 12f
   triplet linear C_4, 11f
Isomerization pathway, linear to cyclic,
   for small carbon clusters, 13
Isomerization reaction, equilibrium
   constant, carbon condensation, 11
Isothermal compressibility, solid C_{60}, 63
Isotopic scrambling experiments to
   determine initial reactants in
   carbon-arc synthesis of fullerenes, 2
Isotopically resolved mass spectra,
   electron-impact ionized ^{13}C-enriched
   C_{60}, 2,3f
Isotopomers of fullerenes, dispersion
   according to binomial distribution, 2–3

K

Kinetic model, collision complex, 5

L

Lanthanum-doped fullerenes, production,
   152,154
Laser shrinking, boronated fullerenes, 149
Laser vaporization
   boron–graphite composite target
     disc, 144
   graphite targets, 153
Lattice parameters, C_{60} and C_{70} pentane
   solvates, 33
Lifetimes, collision complexes, 7–9
Linear carbon clusters
   bending modes and bonding, 10
   formed in condensation of carbon, 5
   harmonic vibrational frequencies, 7–9
   molecular constants, 7–9
   symmetries of HOMOs, 13
   See also Carbon clusters
Liquid-state NMR spectroscopy, C_{60} and
   C_{70}, 108–111

Lithium-doped fullerenes, production, 156
Low-resolution single-crystal X-ray
   structure, fullerenes and monoanions,
   41–51

M

Magnetization
   potassium-doped C_{60}, 79,80f
   rubidium-doped C_{60}, 80,81f
Mass chromatogram, C_{60} osmate esters,
   93,95f
Mass spectrometry
   evidence for small-molecule synthesis of
     C_{60}, 2–4
   fullerenes, 128
   isotopically resolved, C_{60}, 3
   negative ion chemical ionization, C_{60},
     162,164f
   plasma desorption, description of
     technique, 117
   purified sample of C_{60}, 118
Mechanisms
   fragmentation of spheroidal carbon
     shells, 17
   growth of spheroidal carbon shells, 18
Metal atoms, trapped inside fullerene
   clusters, 152
Metal complexes, reaction with C_{60},
   177,180
Metal-containing species, laser-vaporized
   target rods, 154
Microwave loss as function of temperature,
   doped C_{60} films, 79
Molecular conductors, effect of pi
   orbitals, 71
Molecular constants, linear carbon
   clusters, 8t
Molecular parameters, derived from
   measured IR and FIR spectra, 19
Molecular rotations in solid C_{60}, 26
Molecular structure, solid C_{60} from MNDO
   optimization, 82
Monoclinic unit cell, consistent with
   diffraction from pentane-solvated C_{60}
   and C_{70}, 32
Monocyclic rings, formed in condensation
   of carbon, 10
Morpholine, addition to C_{60}, 165
Motion, C_{60} molecules, 30

N

Naphthalene–naphthalenide anion, ESR
spectrum, 42
Negative ion chemical ionization mass
spectrum, C_{60}, 162,164f
Nitrogen-doped fullerenes, unsuccessful
attempts to produce, 151
NMR spectroscopy
liquid state, C_{60} and C_{70}, 108–111
one- and two-dimensional, C_{60} and C_{70},
107–113
potassium-doped C_{60}, 80,81f
solid C_{60}, 56
solid state, behavior and properties of
C_{60}, 111
See also ^{13}C NMR spectroscopy,
Solid-state NMR spectroscopy
Nomenclature, doped fullerenes, 142
Nucleophiles, addition to C_{60}, 162

O

Open-shell synthesis model for
fullerenes, 16
Organic conductors, effect of pi orbitals, 71
Organolithium reagent addition, C_{60},
162–163
Organometallic reagents, reaction with
C_{60}, 178
Orientation, C_{60} molecule, 30
Orientational disorder, C_{60}, and difficulty in
obtaining crystal structure, 92
Orientational order
C_{60}, effect of temperature, 62
C_{60} molecules, 56,61
solvated C_{60}, 46
ORTEP drawing, C_{60}-osmium tetroxide
adduct, 97,98f,99f
Osmate esters, characterization, 92
Osmium tetroxide adduct, bond lengths
compared to C_{60}-Pt complex, 182
Osmylated C_{60}
analysis of crystal structure, 97
bond lengths, 100
histogram of carbon-carbon bond lengths,
102,103f
histogram of distances from calculated
center of C_{60} moiety to each C atom,
99,101f

Osmylated C_{60}—Continued
histogram of sums of C–C–C angles,
99,102f
unit cell, 97,100f
Osmylation
anthracene, 92,94f
C_{60}, 92,94f
Oxidation, temperature-programmed,
fullerenes, 136–137

P

Packing of C_{60} molecules, models, 92,93f
Pentane-solvated C_{60} or C_{70}, model for
twinning, 34
Pentane solvates, C_{60} and C_{70}, 31
Phenyl Grignard reagent, addition
to C_{60}, 162
Pi orbitals, effect of directionality on
electronic properties, 71
Plasma desorption mass spectrometry
description of technique, 117
See also Mass spectrometry
Platinum complexes, reaction with C_{60}, 180
Platinum coordination spheres, C_{60} versus
ethylene complexes, 183
Potassium-doped C_{60}
conductivity, 76
magnetization, 79,80f
microwave loss as function of
temperature, 79
NMR spectrum, 80,81f
resistivity, 76,77f
Preparation, C_{60} and C_{70} anions, 42
Pressure, effect on isothermal
compressibility, C_{60}, 63
Production
gram quantities of fullerenes, 128
macroscopic amounts of C_{60} and C_{70}, 153
Propylamine, addition to C_{60}, 162
Puckered rings, stability, 14
Purification of C_{60}, 167

R

Raman spectra, C_{60} film during rubidium
doping, 77,78f
Reactivity, doped fullerenes, 155
Refinement, based on single-crystal
data, 29

Resistivity
  K_xC_{60} film, 76,77f
  potassium-doped C_{60}, 77,78f
Rovibrational transitions, linear
  odd Cn, 19
Rubidium-doped C_{60}
  magnetization, 80,81f
  microwave loss as function of
    temperature, 79
  Raman spectrum, 77,78f
Ruthenium complexes, reaction with
  C_{60}, 178

S

Scanning electron microscopy,
  pentane-solvated C_{60} crystal, 31,32f
Semiconductors, effect of doping, 141
Separation of fullerenes
  by extraction, 120
  by sublimation, 122
Single-crystal X-ray study, solid C_{60}, 56
Singlet linear C_9, IR spectrum, 12f
Small-molecule synthesis of C_{60}, supported
  by isotopic scrambling and NMR
  experiments, 4
Solid C_{60}, See C_{60}, solid
Solid-state NMR spectroscopy, behavior and
  properties of C_{60}, 111
Solvated C_{60}, single-crystal X-ray
  structure, 41–51
Solvated C_{60} structure, cubic data-set
  refinement, 46
Solvated fullerenes, single-crystal X-ray
  structure, 41–51
Solvates of C_{60}, See C_{60}, solvates
Solvation, problems eliminated with
  sublimation, 29
Spectroscopy, C_{60} and C_{70} anions, 42
Spheroidal carbon shells
  mechanisms of fragmentation, 17
  mechanisms of growth, 18
Spheron, temperature-programmed
  oxidation, 136–137
Structure analysis
  alkali fullerides, 80
  C_{60} carbon cluster, 99
  carbon framework of C_{60}, enabled by
    osmylation, 97
  solvated C_{60}, X-ray studies, 46

Sublimation
  C_{60}, 29
  to separate fullerenes from raw soot, 122
Sublimed crystals, C_{60} and C_{70}, 29
Sublimed films, C_{60} and C_{70}, 123
Superconducting bulk materials, 79
Superconductivity
  alkali-intercalated C_{60}, 64
  alkali metal doped C_{60}, 71–86
Surface analysis by laser ionization
  (SALI), fullerenes, 128,130,132f
Symbolism, doped fullerenes, 142
Symmetry
  C_{60} molecule, 26
  HOMOs of odd-numbered linear Cn
    clusters, 13,14f
Symmetry model, C_{60}, idealized model,
  27–28
Synthesis
  C_{60} from small carbon clusters, 1–20
  carbon-arc, See Carbon-arc synthesis

T

Temperature
  dependence of K_{eq}, chains and monocyclic
    rings, 12
  effect on behavior of C_4 and C_{10}
    clusters, 13
  effect on magnetization of
    potassium-doped C_{60}, 79,80f
  effect on magnetization of
    rubidium-doped C_{60}, 80,81f
  effect on microwave loss, doped C_{60}
    films, 79
  effect on orientational order of C_{60}, 62
  effect on resistivity of potassium-doped
    C_{60}, 77,78f
Temperature-programmed oxidation,
  fullerenes, 136–137
Thermal desorption, fullerenes, 130
Thermal properties, fullerenes, 120
Thermogravimetric analysis, fullerenes,
  120,135
Thin films
  alkali metal doped C_{60} and C_{70},
    conductivity, 76
  C_{60} and C_{70}, preparation and analysis, 73
Three-dimensional clusters, formed in
  condensation of carbon, 16

Three-zone oven, fullerene separation, 123,124f
Titration, boronated fullerene ions, 147–148
Transition metal complexes, reaction with C_{60}, 180
Triethyl phosphite, addition to C_{60}, 165
Triplet linear C_4, antisymmetric stretch, 11f
Twin domains, C_{60} and C_{70} pentane solvates, 33,34f,35f

U

Unit cell
  monoclinic, consistent with diffraction from pentane-solvated C_{60} and C_{70}, 32
  osmylated C_{60}, 97,100f

V

Vibrational state densities and bending modes, 10
Vibrational state density of energized molecule, 7

X

X-ray diffraction
  C_{60} and C_{70} pentane solvates, 31
  C_{60} crystal structure, 27
X-ray powder diffraction profile
  Cs-doped C_{60}, 64,65f
  pure C_{60}, 61,62f
  solid C_{60}, 55
X-ray studies, structure of solvated C_{60}, 46

Copy editing and indexing: Janet S. Dodd
Production: Donna Lucas
Acquisition: A. Maureen Rouhi
Cover design: Robert E. Sargent

Printed and bound by Maple Press, York, PA

Other ACS Books

Chemical Structure Software for Personal Computers
Edited by Daniel E. Meyer, Wendy A. Warr, and Richard A. Love
ACS Professional Reference Book; 107 pp;
clothbound, ISBN 0–8412–1538–3; paperback, ISBN 0–8412–1539–1

Personal Computers for Scientists: A Byte at a Time
By Glenn I. Ouchi
276 pp; clothbound, ISBN 0–8412–1000–4; paperback, ISBN 0–8412–1001–2

Biotechnology and Materials Science: Chemistry for the Future
Edited by Mary L. Good
160 pp; clothbound, ISBN 0–8412–1472–7; paperback, ISBN 0–8412–1473–5

Polymeric Materials: Chemistry for the Future
By Joseph Alper and Gordon L. Nelson
110 pp; clothbound, ISBN 0–8412–1622–3; paperback, ISBN 0–8412–1613–4

The Language of Biotechnology: A Dictionary of Terms
By John M. Walker and Michael Cox
ACS Professional Reference Book; 256 pp;
clothbound, ISBN 0–8412–1489–1; paperback, ISBN 0–8412–1490–5

Cancer: The Outlaw Cell, Second Edition
Edited by Richard E. LaFond
274 pp; clothbound, ISBN 0–8412–1419–0; paperback, ISBN 0–8412–1420–4

Practical Statistics for the Physical Sciences
By Larry L. Havlicek
ACS Professional Reference Book; 198 pp; clothbound; ISBN 0–8412–1453–0

The Basics of Technical Communicating
By B. Edward Cain
ACS Professional Reference Book; 198 pp;
clothbound, ISBN 0–8412–1451–4; paperback, ISBN 0–8412–1452–2

The ACS Style Guide: A Manual for Authors and Editors
Edited by Janet S. Dodd
264 pp; clothbound, ISBN 0–8412–0917–0; paperback, ISBN 0–8412–0943–X

Chemistry and Crime: From Sherlock Holmes to Today's Courtroom
Edited by Samuel M. Gerber
135 pp; clothbound, ISBN 0–8412–0784–4; paperback, ISBN 0–8412–0785–2

For further information and a free catalog of ACS books, contact:
American Chemical Society
Distribution Office, Department 225
1155 16th Street, NW, Washington, DC 20036
Telephone 800–227–5558